NATURALLY SLIM AND POWERFUL

ALSO BY DR. PHILIP LIPETZ

The Good Calorie Diet

Naturally Slim and Powerful

Dr. Philip Lipetz and Monika Pichler

**Andrews McMeel
Publishing**

Kansas City

A hardcover edition of this book was published in 1997 by Andrews McMeel Publishing.

Library of Congress Cataloging-in-Publication Data

Lipetz, Philip.
Naturally slim and powerful / Philip Lipetz and Monika Pichler.
    p. cm.
  ISBN: 0–8362–5284–5 (ppb)
    1. Weight loss. 2. Women—Nutrition. 3. Serotoninergic
mechanisms. 4. Reducing diets. I. Pichler, Monika. II. Title.
RM222.2.L5343 1997
613.2′5′082—dc21   96–39700
                                                           CIP

*This book is dedicated with love to
the goddess in every woman.*

Contents

Foreword *by John Gray, Ph.D.*   ix
Foreword *by William Regelson, M.D.*   xi
Acknowledgments   xv
Author's Note   xvii

PART I: THE FOOD-MIND CONNECTION   1

1. Women and Food   3
2. Serotonin and Dieting   13
3. Serotonin, Food, and a Modern Woman's Life   24
4. A Call to Freedom   40

PART II: THE PROGRAMS   55

5. Creating a Woman-Friendly Nutritional
   Environment   57
6. Good and Bad Calories   79
7. Free-Eating Program   98
8. Caloric Restriction   110
9. Decreasing Stress   129
10. Questions and Answers   137

Appendix I: Good and Bad Calories by Category   154
Appendix II: Glycemic Index of Selected Foods   168
Appendix III: Mail-Order Resources   177
Appendix IV: Recipes   179

Notes   247

Foreword

BY JOHN GRAY, PH. D.

Men Are from Mars, Women Are from Venus shows how men and women differ in all areas of their lives. Not only do men and women communicate differently but they think, feel, perceive, react, respond, love, need, and appreciate differently. They seem to be from different planets, speaking different languages and needing different kinds of nourishment. It turns out that the different kinds of nourishment men and women need is not only emotional, but dietary as well.

Naturally Slim and Powerful reveals the special nutritional needs of a woman's body. It is a manual for losing weight while enhancing a woman's well-being. This book does for female physiology and dieting what *Men Are from Mars, Women Are from Venus* did for understanding a woman's emotions. It is an incredibly effective method of losing weight that utilizes the hidden power in a woman's brain.

Our society sends the message that women must control themselves when it comes to food. It expects women to starve their bodies and resist their impulses to eat and feel satisfied. Not only does this approach make women feel weak, it actually wreaks havoc with their brain chemistry, forcing women to fight a battle they cannot win. Irresistible impulses to eat are nature's way of restoring a woman's serotonin back into a state of balance.

Many of today's foods and diets actually rob women of control over their lives. Our lack of understanding about women's unique nutritional needs is why the vast majority of prescriptions for serotonin-stimulating drugs such as Prozac, Redux, and fenfluramine are given to women. Unfortunately, this creates a state of dependency and does not acknowledge the true strength of women.

True control happens naturally. The feeling of being in control really flows from not being consciously in control. If you have to apply willpower it means you are fighting something and therefore do not really have control. *Naturally Slim and Powerful* shows women how to be in control without engaging in a constant battle with food. Well-being flows from using foods that meet a woman's unique needs, creating a totally new relationship to dieting.

Women can lose weight and keep it off when they eat foods that support a woman's unique brain chemistry. The very foods that work best for men can actually prevent women from looking and feeling their best. The ideal diet for a woman not only allows her to shed unwanted pounds but raises serotonin levels, which in turn enriches her emotional life. With a balanced diet designed specifically for her unique physiological needs, a woman can begin to enjoy again the freedom of eating what tastes good without starving herself again and again.

No longer is dieting an exercise in willpower. Women feel in charge of their lives. Instead of saying to themselves, "What's wrong with me, why can't I control my eating?" they find themselves losing weight easily and feeling really good.

I am personally very excited to apply these ideas and share them not only with women but with the men who love and want to support them.

Foreword

BY WILLIAM REGELSON, M.D.

The publication of *Naturally Slim and Powerful* occurs at a critical moment, when *Time* magazine and other popular publications call attention to the public that obesity and uncontrollable weight gain may be a metabolic disease. The accumulation of excess body fat affecting more than two thirds of all Americans is not just an act of slothful, sinful overindulgence, but a physiologic event that we are just learning to understand. Obesity is now seen as categorizing energy pathways and behavior, and those of us who suffer from this problem are no more guilty than the diabetic whose insulin deficiencies are intrinsic to genetic and evolutionary survival mechanisms.

Philip Lipetz, in his first book, *The Good Calorie Diet* (Harper-Collins, 1994), presented the overwhelming evidence that excessive weight gain was a metabolic problem related to the episodic starvation that characterized early human evolutionary development. Under the marginal and primitive conditions related to our appearance as a species on Earth, we had to store up fat in order to survive the lean starvation periods of our existence. Living off stored fat enabled us to survive and spread our DNA-containing progeny throughout the world. It helped us to survive as a species; however, it is also a mechanism that makes us fat.

It turns out that the hormone insulin plays a key role in this. It regulates our energy reserves, lowers blood sugar, and is critical to maintenance of muscle mass and fat storage. *The Good Calorie Diet* spoke of the value of cross-linked starch and sugar storage as the mechanism for controlling fat deposition. The book talked of the need for carbohydrate foods that would slowly be digested to release sugar-related calories, preventing excess production of insulin, which can convert sugar to its stored place as fat energy. Philip's diet concept was to provide selected complex, cross-linked starches that would be digested slowly to avoid rapid sugar highs and lows and the fat-inducing effects of insulin released by too rapid a load of sugar. *The Good Calorie Diet* presented the need for the slow, constant release of glucose from proper food sources to avoid rapid shifts in blood sugar. What is exciting is that not all carbohydrate-containing foods are the same. One can easily identify those foods that do not encourage fat formation, which are filled with what Lipetz called "Good Calories" and those that increase fat storage filled with "Bad Calories."

Now, in this groundbreaking new book, Lipetz and his coauthor wife, Monika Pichler, add another dimension to preventing obesity. They show the association between mood and behavior as governed by serotonin synthesis, which affects mood, appetite, and fat production. They say that there is a food-mind connection that equates mood with the desire for food. This is most important to women and of less concern for men. The fact that women have a harder time controlling their weight than men is well known. The explanation for sex differences is that serotonin is more important to a woman's sexual role than a man's. She needs to maintain brain serotonin levels to contend with child rearing and the stress of her dual role as food gatherer and mother.

As seen in recent articles in the popular press, drugs that affect serotonin levels, such as Prozac, Redux, and Pondimin, can, at least for several months, decrease food intake and fat deposition. The problem with starvation or low-calorie dieting, however, is that they stimulate both insulin increases and serotonin depletion, which makes you irritable and depressed, and which causes food cravings. This kind of dieting can set up a futile cycle that forces you to store fat despite restricted calories.

How many of us can sustain food withdrawal without irritability, depression, and relapse? Our addiction to food is a survival-mediated event, and it is impossible to do away with it effectively, but what Lipetz and Pichler have determined is that certain foods—that is, complex carbohydrates slow to release their sugar—help us to lose weight. These complex sugars and starches also help to restore our levels of brain serotonin, which relieves us of diet-induced anxiety, food cravings, and depression. These complex carbohydrate foods, outlined in this book, are like those familiar "spansule capsules" that slowly release their "sugar" content over time.

Alternatively, in addition to dieting as outlined in this book, we can go on serotonin-enhancing drugs like Prozac, Redux, or fenfluramine, but their effectiveness is only measured in days or weeks. For this reason, one has to think of using the programs listed in this book in conjunction with taking these pills. These diets will not only permit prolonged and comfortable dieting, but could act with any serotonin-enhancing drug to decrease appetite and mobilize fat, aiding in extending the value of any diet program or diet pill.

Additionally, with the growing popularity of women's participation in sports, athletic coaches and female athletes should be aware that training diets should differ between men and women. It should not be assumed that a man's training diet is appropriate for a female athlete. Similarly, we should be concerned with hospital and institutional nutrition that does not recognize the special nature of a woman's nutritional needs.

Although this diet book stresses the differences between the responses of men and women, it should be noted that as men grow older many of the endocrine differences related to the male hormones decline, which has biologically, to some extent, a feminizing effect on men, making them candidates for the program outlined in this book.

In addition, serotonin depletion is responsible for nausea and vomiting following chemotherapy and radiation. It can also make many of us oversensitive to drugs, certain foods, and our environment, conditions that can all be ameliorated by the *Naturally Slim and Powerful* program.

In conclusion, women must be recognized to be metabolically different than men, and for this reason it is not surprising that their dietary strategies to lose weight and mobilize fat should have a different pattern of clinical sensitivity, leading to differing drug utilization and diets. In this case we must avoid political politeness and shout "vive la difference!" as we pay true respect to weight-controlling strategies that must be distinct between men and women.

The strategies outlined in this book will reward us with healthier bodies and minds, particularly for women for whom weight control is a mortifying and almost insolvable problem. This breakthrough program succeeds where conventional diets fail. It recognizes that all women, not just dieters, need the right foods to sustain their unique serotonin balance.

Acknowledgments

We wish to thank the following individuals who contributed to this book: Rebekah Lipetz, Dotti Lipetz, Margrit Pichler, Sheela Dunovan, and Carol Mann, our agent.

Most of all we would like to thank the hundreds of people who wrote and told us of their experiences using *The Good Calorie Diet*. Their enthusiasm and results are what motivated us to expand this program and discover a woman-friendly nutritional environment. In particular, we would like to thank the women whose quotes are used in this volume. All of the quotes in this book are real and represent the real experiences of women.

This book was written using the IBM VoiceType dictation system, which allows your spoken words to automatically appear on your computer screen. This is but the first of many systems that in the next decade will totally transform the way we interact with computers, making information and power easily available to anyone who can speak.

Authors' Note

As with any significant change in diet, you should undertake our programs only after consulting with your personal physician or other health professional. Although our research has led us to conclude that this diet will help you lose weight while protecting your serotonin, you should not attempt to use this program as a substitute for any condition that requires the use of serotonin-stimulating medications. We also believe this program will help control insulin; however, you should never use this program as a substitute for a medically prescribed program of diabetes control.

Any program designed to control insulin can have interactions with medications; therefore, if you are taking any prescribed medications you should consult with your physician before undertaking this program. If you suffer from diabetes; clinical depression; psychiatric disease; hypoglycemia; gallbladder, liver, kidney, digestive, or other chronic disease; or if you are pregnant, you should not follow our program except under a doctor's close supervision.

If you are currently taking serotonin-stimulating medications you must consult with your prescribing physician before undertaking this program. Several clinical studies show the efficacy of these medications changes as your diet changes. When you switch to a diet that supports serotonin synthesis you may be inadvertently increasing the power of your medications, resulting in overdosage. It is also possible this program may make you feel so good you may wish to change your prescribed dosage. Never make changes in any prescribed medication without consulting with your physician.

Part I

THE FOOD-MIND CONNECTION

CHAPTER 1

Women and Food

Women have long known what modern medicine has just discovered. Any woman will tell you her body responds differently to food than a man's. She gains weight more easily, loses weight more slowly, and has food cravings and obsessions rarely seen in men. Her reaction to food is more physically responsive and emotionally charged.

A startling new discovery explains why women are so different from men when it comes to food and will forever change the way we think about dieting and a woman's nutritional needs.

The key to this discovery is the brain chemical serotonin and how it affects women's well-being. At some point in her life, the average woman eats the wrong foods or goes on a diet, and the resulting serotonin deficiency shuts off important parts of her brain. Low serotonin is the reason she gains weight and has more food cravings, mood swings, and many of the problems that preferentially afflict women.

We are going to have to change the way we think about women's physical, emotional, and mental health. Women are much greater than their food allows them to be. None of the symptoms of low serotonin are an inherent part of being a woman.

It is possible to have it all—a slim body, a powerful mind, and freedom from many "women's complaints." All a woman has to do is eat foods that sustain her serotonin chemistry.

Our society's reliance on weight-loss diets designed for the male body previously forced women to choose between three unacceptable alternatives. One, they could continue on male-oriented diets but would suffer from the many problems brought on by low serotonin. All conventional diets lower a woman's serotonin. High-protein, low-fat, and high-carbohydrate diets all bring on abnormal serotonin. Even conventional serotonin-boosting diets fail to meet women's special needs, and may lower serotonin.

Two, they could choose an increasingly popular alternative and continue with male-oriented diets and take serotonin-stimulating drugs such as Prozac, Redux, or fenfluramine (fen/phen). However, these drugs are expensive and carry the risk of many harmful, and even fatal, side effects.

Or three, they could choose what was previously the only non-pharmaceutical way to support serotonin—abandon dieting. The foods that women turn to when they leave their diet are almost always fattening foods that restore serotonin levels to normal.

This meant that as long as women continued to ignore their special needs there was no nonpharmaceutical way to both diet and support serotonin.

We will show women how to eat so that they support their serotonin. Instead of drugs, we offer two programs that support a woman's serotonin, increase her well-being, and allow her to lose weight easily.

SEROTONIN AND A WOMAN'S PSYCHE

Women need different foods because their brains have much more serotonin than men.[1] Maria Carlsson and her colleagues at the University of Gothenburg, in Sweden, found that "the serotonin neurons of the female . . . brain have a greater storage capacity, a higher enzymatic activity with a higher rate of serotonin synthesis and are thus generally more developed than in the male."[2] This difference arises because female sex hormones enhance serotonin activity, while male hormones inhibit serotonin.[3]

A woman's higher levels of serotonin may be of fundamental

importance for such functions as appetite, sexuality, impulsive behavior, and aggression.[4] Serotonin also controls the way women eat, drink, and even seek pleasure.[5] Serotonin is like a surrogate parent, discouraging negative actions and comforting us during hard times. Serotonin is what gives women the power to better withstand stress and be more nurturing, serene, and peaceful than men. Women are not as subject to impulsive behaviors. They're more cooperative. They're more intuitive. Simply put, serotonin makes women different from men.

Shirley MacLaine describes these unique feminine qualities very aptly in her Hollywood memoir *My Lucky Stars*. She recalls how different it was working with a predominantly all-women cast in *Steel Magnolias*. The actresses were Sally Field, Dolly Parton, Olympia Dukakis, Daryl Hannah, and Julia Roberts.

> So our gang of wonder women met, worked, and lived together. We cried, laughed, and teased each other. I don't remember a moment of jealousy, envy, or proprietary behavior. In fact, each of us was more concerned for the others than we were for ourselves.
>
> We knew we were part of a new feminine sensibility that was as efficient as that of men, but operated with a compassion and intuition that was much more effective. The crew noticed it right away. At first they wondered if we'd disintegrate from within, running afoul of the usual creative conflicts and differences. When we didn't, they began to truly study our ways. They saw how we came to each other's aid when one of us was in trouble with a scene. Sometimes we'd ask the director to leave us alone while we collectively rushed in to help our own. We covered for each other, we cooked for each other, we joked with each other, and we respected each other's privacy. It was an experience not unlike what people saw on the screen when the movie came out, only we weren't just in character, we were being ourselves.

Women with low serotonin cannot experience a sense of well-being. Low serotonin creates a sense of panic, anxiety, and depression. Without the help of serotonin, small problems can become large. Think of serotonin as being the police force that prevents a small disturbance from escalating into a riot.

At some time, almost every woman has suffered from some preventable malady brought on by low serotonin. It affects every woman differently. Some women know they are ill. Others think their infirmity is just a normal part of living, the results of age, or just another "women's complaint" that must be endured. The vast majority are not even aware that low levels of stress are not an inherent part of modern life but rather symptoms of a nutritional disease.

Symptoms of low serotonin include:

♦ weight gain
♦ depression
♦ stress
♦ tendency toward substance abuse
♦ PMS
♦ food cravings and eating disorders
♦ sexual dysfunction
♦ anxiety
♦ irritability
♦ disturbed sleep
♦ headaches
♦ obsessive and compulsive behavior
♦ inability to form proper social relationships
♦ lower social dominance

You may be skeptical that serotonin could be the solution to so many problems afflicting modern women, but a more detailed review (see Chapter 3) should convince you of serotonin's remarkable effect on a woman's health.

SUPPORTING A WOMAN'S SEROTONIN

Because women have more serotonin, they have a greater need to eat foods that support serotonin synthesis.[6] This means that women are more sensitive to any decrease in serotonin-stimulating foods.[7]

Without these foods, serotonin levels will fall, making it impossible to use all the potential power nature designed into every woman's brain.

If men do not eat serotonin-stimulating foods, they will notice very little change in their behavior.

Nature endowed all women with the capacity to have slim, healthy bodies and the ability to function powerfully at home and at work. Yet, low serotonin creates an entirely different picture of women. The same food that nourishes and sustains men can lower a woman's serotonin and attack the core of her physical, emotional, and mental well-being.

We became aware of how many women suffer from the effects of the wrong food when we started to receive countless letters from women describing the miraculous results they experienced with *The Good Calorie Diet*, Philip's previous book. Thousands of female and male readers lost weight, but women reported that the diet elevated their mood, made their food cravings vanish, and lifted their depression.

> What amazes me is how good I feel. All of my food cravings are gone. I no longer am afraid to sit down to eat. I also feel so much happier, it is as if a veil of depression has been lifted from my life. Instead of living in fear, I now want to go out of the house and do things. My husband is thrilled, both with my weight loss and with my new spirit.
>
> Again, thank you for offering something that really works.
>
> —*R. K., New York City*

> What was most amazing was that I no longer felt constantly depressed. Now I feel good all the time. I felt a permanent psychological shift since starting the diet. My mind is clearer, I can think better, I remember more, and I feel more confident now when expressing my opinions. For the first time in my life, I feel sexy. Truthfully, I've never enjoyed sex so much. Thank you. Thank you. Thank you. I feel like I was set free!
>
> —*S. S., Oakland, CA*

> After the second week, I felt like a different person. I no longer craved fattening snacks at night. I no longer had headaches or

backaches and my endurance during exercise almost tripled! Best of all, I'm in a good mood all of the time now!!

—*L. P., Seffner, FL*

I received your book from a doctor who specializes in weight control. After trying your theory, I have lost twenty-two pounds. It is amazing. I have a history of migraines since I was two years old. I have not had one headache eating this way.

—*C. V., Visalia, CA*

I am thin and went on your program to help me maintain my weight and stop bingeing. It did just that! I no longer crave sweets and [now can] eat "enormous" amounts of food. I don't weigh myself anymore but I remain thin. I am a normal person.

—*S. B., Bayside, NY*

I like the fact that I don't have to starve and feel hungry all the time. I don't have the cravings I use to have. I don't have to deny myself or be depressed because it actually feels effortless. I never miss a meal, I eat snacks and I rarely exercise. Some people who haven't seen me for a few months don't recognize me. Others remark that I look ten years younger. I use your book like a second bible, when it comes to eating. I don't plan to ever be without it. I am so delighted in being able to lose weight and keep it off without agonizing over it. I actually feel as if my body is normalizing. With the exception of surgery no other weight loss method has helped me like this. Even surgery didn't help me keep it off like this.

—*W. B., Wenatchee, WA*

As we read these stories, we saw example after example of what women have always known, and modern medicine is just discovering—that women have completely unique physical, mental, and emotional responses to food.

These women experienced a change in both their bodies and their minds. Every one of these positive changes—reduced food cravings and appetite, weight loss, elimination of headaches, increased sexuality, relief of depression, increased confidence, reduced fatigue, less fear and anxiety—is the simple result of a diet that restores a woman's brain chemistry to normal, to its fullest potential.

Losing weight is an added bonus of a diet that restores a woman's serotonin. You lose weight because a woman-friendly diet balances the abnormal brain chemistry that forces women to overeat and abandon diets. You lose weight because this diet, unlike any other diet, stimulates serotonin while also reducing insulin, the prime fat-forming hormone.

ONLY WOMAN-FRIENDLY DIETS WORK

Many problems vanish when women learn how to feed and care for the special needs of the female brain, and woman-friendly foods do just that. Having enough serotonin gives all women the opportunity to feel happy, calm, full of energy, sexually alive, and easily able to maintain their optimal weight.

This is not a book of psychological tricks. Women do not need to learn how to compensate for shortcomings. Women have no inherent shortcomings—just improper nutrition.

Our serotonin-boosting program has been proven in a major study published in *The American Journal of Clinical Nutrition* to produce up to 65 percent more weight loss than conventional diets. The special property of our program is the way it elevates serotonin.

Our program is very different from other serotonin-boosting diets, which use male-centered carbohydrates to produce insulin surges. Every other diet, except these serotonin-boosting diets, goes out of its way to lower, not raise, insulin surges, which stimulate appetite and fat formation.[8] Furthermore, conventional serotonin-stimulating diets make women susceptible to all forms of addiction—overeating, as well as the abuse of tobacco, alcohol, and drugs (see Chapter 3).

Our program works better simply because it is the only serotonin-boosting program to elevate serotonin without also producing excess insulin. The same clinical study that proved woman-friendly carbohydrates produced more weight loss also showed that these foods control the insulin surges that produce food addiction and excess fat formation.[9]

When you experience the magic of woman-friendly carbohydrates you will be like this woman who lost weight, felt great, and increased her brain power.

When I first tried your program I thought you were a trifle bonkers. A calorie was a calorie, and all I needed to do was to eat less of them. Preferably one leaf of arugula a day, lightly tossed with some Evian water. This method of weight loss and maintenance didn't leave me feeling all that great, but what else was there? I felt about as friendly and relaxed as the Tasmanian devil, and I felt constantly upset. My body ran my mind and there was no stopping.

The miracle came shortly after starting the program. One day, I was reading and after a few pages I looked up, noticing something was quite different. I slowly realized not only was I not eating, but I had no desire to eat. I was not planning what to eat, where to get it, when to eat it, as I normally do. I was not obsessing about food. For the first time, I felt contented and peaceful inside my own body.

I could feel a complete permanent psychological shift in my body. I felt happiness and joy—a quiet feeling of elation. My mind had become very quiet, it was no longer being run by my body.

Over the next two months, I lost over ten pounds without thinking about it! It was easy because I no longer had cravings for chocolate or fats. My appetite also seemed to go away.

But what was most amazing was that I was no longer constantly depressed. Previously, the only time I felt good was when I was eating. Now I feel good all the time. The feeling of a clear, quiet mind has continued, making it possible for me to go shopping, out to dinner, or to sit with friends without the constant obsession of food on my mind.

—*S. D., Oakland, CA*

Finally, a woman can have it all—weight loss and a healthy brain. What's more, this is not a program of extremes; 50 percent of the calories are from woman-friendly carbohydrates, 25 to 30 percent from fat, and 20 to 25 percent from animal or vegetable protein.

We find over and over that women who eat the right foods, especially while dieting, experience transformations in their minds as well as their bodies.

It also explains why our program produced such extraordinary results for women.

AN APPEAL FOR WOMEN

It is time for the medical profession to realize there is a significant correlation between the problems induced by low serotonin and the things that seem to preferentially afflict today's women. We say much of the problem is the food they eat.

Such an explanation is compelling because it suggests that treatment is just a matter of substituting proper foods for improper foods and traditional weight-loss diets. It also suggests that women have a lot more control over their lives, something we've seen to be true in women who've tried our programs. When women start eating diets fashioned for their specific needs and stop eating meals tailored for men, they experience profound life changes, both physically and emotionally.

This book takes the first step toward establishing a nutritional basis for many of the problems specific to women.

For many women the idea of being different from men is a sensitive topic after fighting so hard for equality. Difference is associated with better and worse, inferior and superior. In no way do we want any of this book to be used as a biological basis for discriminating against women. In fact, it can be argued that their unique serotonin chemistry makes women superior to men. But all of these arguments are irrelevant. Our aim is to educate women about their nutritional needs, which are ignored by today's society.

Our society has a great fear of claims of biological differences. It wasn't that long ago that Nazis used bogus biological claims to justify genocide. American racists defend their segregationist policies with claims that African Americans suffer from inferior intelligence.

However, gender difference is a fact. People are taller or shorter, have more or less muscle mass, their eyesight is good or bad. The question is not whether differences exist but whether we use differences to discriminate or to remove inadvertent discrimination.

We do not have to translate differences between the genders into lesser or greater opportunity, into inferior and superior categories. If modern food harms women and not men, then acknowledging and curing that problem is a beneficial use of biological differences. Continuing to ignore the nutritional needs of women will just deny them the opportunity to have slim bodies and powerful minds.

What is most relevant is whether the proper foods will change the lives of women. We have found that women's lives are profoundly affected by eating the proper foods.

In order for you to experience how foods can act like mind-altering drugs, stay with us for the next two chapters. If you do, your entire idea of women's nutrition will change. Along the way, it will be important to release the ideas created by a food industry that fails to produce lasting weight loss in 98 percent of dieters.

However, you should be aware that serotonin problems are confined to dieters; the same gender-insensitive ideas of nutrition blind us to the nutritional needs of all women. In the conclusion to this book we will see that all women suffer unless they eat woman-friendly diets.

CHAPTER 2

Serotonin and Dieting

Dr. Philip Cowen's group at Oxford University in England has spent the last decade researching how serotonin alters our behavior. Unlike most researchers, he pays particular attention to differences between women and men. His work is the best proof that conventional weight-loss programs damage a woman's, but not a man's, serotonin.

He looked at tryptophan, the amino acid that is made into serotonin. Less tryptophan enters the brain to be made into serotonin when a woman goes on a diet. Without new serotonin, she soon uses up her old serotonin and becomes susceptible to any of the conditions that accompany low serotonin levels—principally depression, food cravings, irritability, and decreased well-being.

Dr. Cowen's research group showed that three weeks of a low-calorie diet (1,000 calories for women and 1,200 calories for men) brought on low serotonin in women but not in men.[1] He has repeated this experiment several times and always found that a dieting woman *always* suffers from greater changes to her serotonin system.[2]

This difference between the sexes is a proven fact. Researchers at Yale University and in Canada, Spain, and Germany confirm that dieting deprives the female brain of the capacity to maintain adequate levels of serotonin.[3] Other research groups confirm that dieting does nothing significant to the serotonin-making capacity of men.[4]

Low serotonin creates depression and carbohydrate cravings in women.[5] Therefore, it is not surprising that standard biochemical tests of depression frequently show depression in dieting women, but not dieting men.[6]

It does not take long for these effects to take place; just three weeks of a low-calorie diet creates depression.[7] As little as six-and-a-half pounds of weight loss was enough to make some women depressed.[8]

Fortunately, not every dieting woman undergoes a depressive crisis. However, even a small depression can make a dieting woman feel badly and reduce her effectiveness in the world.

Low serotonin creates more problems than depression. Dr. Cowen has said that "dieting decreases the availability of [tryptophan] to the [female] brain [which creates] a tendency to binge eat, poor impulse control and depressive symptomatology."[9] He goes on to suggest dieting creates a long list of serotonin-related problems in women:

♦ increased appetite
♦ preoccupation with food
♦ binge eating
♦ depression
♦ irritability
♦ sleep disorders
♦ impulsive behavior
♦ eating disorders, such as anorexia and bulimia

In many ways it is only a matter of chance which symptom of low serotonin strikes a woman after she diets. It can be any of these conditions or the many other serotonin-related conditions listed in the next chapter. One woman may be depressed and suffer from food cravings. Another may be irritable. The individual symptoms will vary, but one thing will always remain constant—low serotonin from dieting decreases a woman's sense of well-being.

Studies show women who go on semi-starvation, weight-loss diets for six months have so many problems with mood, lethargy, depression, irritability, and decreased feelings of energy that it is difficult for them to function effectively.[10]

Unfortunately, it does not take much of a diet to impair a woman's sense of well-being. Dieting just a little—eating about 1,575 calories per day—can result in impaired vigilance, poorer immediate memory, and longer reaction times.[11]

If a woman doesn't diet but just "watches her weight," leaving the table while only a little hungry (1,800 calories per day), she will still not be able to think as clearly as she could when she ate fully.[12] Unfortunately, most modern women "watch their weight" and open themselves up to the risk of low serotonin.

OVERWEIGHT WOMEN SUFFER MORE

Women who are significantly overweight have less serotonin and make less serotonin, even before they go on a diet.[13] Perhaps low serotonin explains why more than three quarters of obese women suffer from depressed moods of short duration, accompanied by carbohydrate cravings and weight gain.[14]

This means that dieting is more difficult for obese women. Imagine how easy it is for a conventional weight-loss diet to further lower serotonin if a woman already suffers from impaired serotonin-making capacity.

SEROTONIN, INSULIN, AND FOOD ADDICTION

Insulin controls serotonin synthesis. Tryptophan, the amino acid that is made into serotonin, is ordinarily stored in reserves that are not available for serotonin synthesis. Insulin causes stored tryptophan to be released and transported into the brain where it is converted into serotonin.

Insulin's primary function is to control the blood sugar created when you digest carbohydrates. Therefore, eating carbohydrates elevates serotonin. This is why so many dieting women with low serotonin develop uncontrollable carbohydrate cravings.

Carbohydrates are so effective at increasing serotonin that some diets suggest insulin-boosting carbohydrates such as chocolate, sweets, bread, sugars, and starches. In the short term, these diets work because these insulin-producing carbohydrates can increase serotonin synthesis by up to four times.[15]

Unfortunately, excess insulin also promotes fat formation. This is why many popular diets, such as high-protein diets, strive to lower insulin.

Too high a level of insulin also changes a woman's sense of taste and makes sweet foods taste better.[16] High insulin also increases her appetite and makes her eat too much, increasing the amount of food eaten in the next meal by more than 50 percent.

As if that weren't enough to warn us away from diets that produce excess insulin, it turns out that high insulin produces only short-lived increases in serotonin. About six hours after insulin levels go up, the body reacts to having too much insulin. The result is a decrease in serotonin synthesis.[17]

It appears that excess insulin also produces excess serotonin, which ultimately decreases insulin to sometimes dangerously low levels.[18] Therefore, creating high levels of insulin to stimulate serotonin just produces temporary relief and ultimately leaves you feeling worse than ever, needing to eat more carbohydrates to boost low insulin and the low serotonin that follows low insulin. You should seek only enough serotonin to give relief, and you should avoid excess levels of insulin and serotonin, which will ultimately produce problems.

You know that feeling of euphoria you experience when you're depressed and eat a candy bar? That is the rush from elevating serotonin. Unfortunately, eating high-sugar foods such as candy bars elevates insulin excessively and ultimately depletes serotonin. High insulin also changes your appetite and sense of taste. It creates a delayed desire to eat more sweets and fats.

This is the classic cycle of addiction. Dieters are caught in a vicious cycle in which they must constantly use their "drugs"—carbohydrates—to relieve relentless cravings and depression produced by high insulin and the resulting fall in serotonin. The only difference is that food addiction is called carbohydrate craving, thus making it seem to be something more controllable.

More and more doctors are becoming aware that dieting women can literally become addicted to serotonin-stimulating fats and sweet foods. The *New York Times* of February 22, 1995, quoted Dr. Louis Aronne, an expert on obesity at New York Hospital–Cornell

Medical Center, as saying, "Can we have an addiction to food? The answer is yes, absolutely."

CONVENTIONAL DIETING AGGRAVATES OTHER ADDICTIONS

Food is not the only thing that elevates serotonin; tobacco, drugs, and alcohol have the same effect.[19] Thus a dieting woman's need for serotonin makes her seek anything, including addictive substances, that restores depressed serotonin to normal levels.

The correlation between female dieting and addiction is best illustrated in the work of Dr. Adam Drenowski of the University of Michigan. Although he is not the only (or even the first) person to investigate this correlation, we are impressed by his work because his study addressed women as a separate group.

An article in *The Journal of Substance Abuse* reports on a study of 1,796 women entering their freshman year in college.[20] Dr. Drenowski compared each woman's severity of dieting to her degree of drug and alcohol use. He found that as the severity of dieting increased, so did the tendency to use alcohol or drugs.

More and more doctors are coming to realize that postdieting substance abuse is just another face of low serotonin. The same serotonin-enhancing drugs that can eliminate carbohydrate cravings can also be used to treat addictions to heroin, morphine, amphetamine, cocaine, tobacco, and alcohol.[21]

The serotonin-addiction connection is so strong that some researchers claim the same serotonin gene underlies food, alcohol, and drug addiction. Dr. Drenowski's work confirms that postdieting preferences for addictive substances involves the same parts of the brain that respond to eating carbohydrates.[22]

MALE-CENTERED DIETS

Clearly, conventional high-carbohydrate diets that seek to increase insulin are not woman-friendly, despite the fact that they elevate serotonin. They rely on male-centered foods that adversely affect the female serotonin system while doing nothing harmful to a man's.

Conventional diets can also be male-centered when they seek to lower insulin. All high-protein diets, such as the Atkins Diet and The Zone, work by inhibiting insulin. These diets are extremely effective at inhibiting fat formation and will produce excellent weight loss in the first few weeks.

The problem with insulin-inhibiting diets is they do not produce enough insulin to stimulate the high levels of serotonin synthesis needed to sustain a woman's needs.[23] Within a few weeks, female dieters find themselves afflicted with the symptoms of low serotonin.

Once again, this problem is much worse in women than in men. With a lesser need for constant serotonin replenishment, men are not adversely affected by these diets. In many ways, a high-protein diet is the ultimate male-centered diet. Only men will lose weight without disturbing their brain chemistry.

When a woman goes on a high-protein weight-loss diet, her brain gradually becomes starved for serotonin. It won't happen quickly. She will notice nothing until she burns up her old stores of serotonin.

The first stages of a high-protein diet produce dramatic weight loss because high protein also inhibits the insulin needed to stimulate fat formation. After a few weeks, serotonin falls past a critical level and the resulting problems force a woman to abandon her diet. This is why high-protein diets produce such great initial results but ultimately create such cravings that most women abandon them.

Low-fat diets may also impair serotonin. It is clear that the brain cannot function properly unless the proper fats are ingested. There is preliminary evidence that an extremely low-fat diet (13 percent of calories or less from fat) interferes with serotonin function.[24]

There is another problem with extremely low-fat diets. Doctors at Rockefeller University suggest that the body makes up for any deficiency in fat by creating new fat from carbohydrates whenever fat levels drop below 20 percent.[25] This was exactly what Philip predicted in *The Good Calorie Diet*. Clearly, any weight-loss diet should not create new fat-forming pathways.

Therefore, a woman should avoid high-protein diets, conventional high-carbohydrate diets, and extremely low-fat diets.

Women will require something new, something centered around their special needs.

WHY WOMEN ARE SO DIFFERENT

Why are a woman's serotonin needs so different from a man's? Why do only women respond to weight loss with altered serotonin? Far from being an inherent flaw, this mechanism insured the survival of the human species.

Our ancestors never lost weight for aesthetic reasons. If they lost weight it was a sign that famine threatened their survival. In such a case it was crucial for women, as caretakers of the young, to survive. Men were less important. In fact, from an evolutionary perspective it makes sense to have some men die because that creates more food for the surviving women. One man can father an entire generation of children, but a woman cannot simultaneously nurse more than a few children. Thus more women than men are needed to create the next generation and care for the current generation.

Therefore women, and only women, developed a complicated set of reflexes to the threat of starvation. Changes in serotonin are but one part of this reflex. Called the Starvation Response, modern women still carry this reflex. Triggered by changes in food and starvation, these reflexes are why modern women have a unique response to food.

The Good Calorie Diet shows how the Starvation Response turns a woman into a storehouse of fat so she and her nursing children can survive the hard times to come. This is similar to the hibernation reflex that fattens bears to survive winter starvation and then sleep through the winter.

The Good Calorie Diet focuses on how this reflex increased the conversion of carbohydrates into fat during times of starvation. It reviews studies indicating that the hormones and enzymes involved in fat formation could become supercharged, increasing over tenfold in activity. The easiest way to describe this change in fat formation is to say that when the body fears starvation it treats every calorie of food as something precious. Everything not needed for the moment should be stored as fat, not excreted.

We will not detail the changes in fat formation that accompany the threat of starvation; for that see *The Good Calorie Diet*. Instead, we will focus on the changes in serotonin that drive a woman to depression and overeating.

Low serotonin is part of the survival strategy unleashed by the Starvation Response. A woman will be able to survive a winter's starvation and feed her young if low serotonin increases appetite and food cravings, accelerating the production of fat.

If nature's survival strategy is to make a woman into a fat storehouse, it's perfectly logical for her behavior to change so she moves around as little as possible; thus exercise does not burn away fat. Low serotonin does this by creating depression. A depressed woman does not move around as much.

All of this insures the survival of the young. Because a depressed mother burns less food energy, she has more calories left over to create milk for her young. Depression also makes women more likely to care for their young because nursing and nurturing children alters brain chemistry and relieves depression. This mechanism favoring the feeding of babies is so perfect that low serotonin even alters critical hormones to increase the production of milk and fat.[26]

Because any additional children would place an intolerable burden on the mother, low serotonin also decreases a woman's sex drive.[27] To further insure there are no new babies during starvation, low serotonin also inhibits normal thyroid function to decrease fertility. Thus, low serotonin insures the survival of the current crop of children, protecting them from having to compete with other babies for food at times when there is not enough food to feed more mouths.

THE STARVATION RESPONSE IN MODERN LIFE

This reflex is still at work in modern women. When they go on a weight-loss diet, serotonin levels fall, and this reflex forces women into abnormal behaviors designed to maximize their weight and sustain their ability to act as the food source for their babies.

Here is how the Starvation Response works in modern times. When a woman starts a diet she expects to experience some fatigue, food cravings, and mental discomfort. She probably considers these problems an inevitable price to pay for a slim body. However, these problems intensify as she continues to deplete her serotonin.

After a few weeks or a few months, the serotonin problems become intolerable. The dieter falls into a minor depression. Food cravings become obsessive, so she gives in and cheats on her diet. This immediately elevates her serotonin, making her mind more peaceful and content. Thus, she is rewarded for overeating. So she cheats on her diet again. The cycle of cheating and reward continues until she has totally abandoned her diet and regained her lost weight.

She decides to favor her mental health over losing weight. Of course, when a woman cheats on her diet she does not say, "I am going to feed my serotonin by eating these fattening foods." She instinctively knows those foods make her feel better. Never does she realize that an ancient reflex has ensnared her.

The Starvation Response forces a dieting woman to alternate between losing weight and having healthy serotonin levels. Because the Starvation Response is unleashed by weight loss, it vanishes when a woman starts overeating. This is why women have been forced to choose between losing weight while suffering from low serotonin or giving up dieting and feeding their serotonin needs.

Unfortunately, modern women are fighting a battle with an opponent so powerful it insured the continuation of the human species. No wonder it's such a struggle to lose weight! This is why over 98 percent of all diets fail.

We hope this helps women lighten up and stop being so hard on themselves. Until they take care of their special needs there is no way women can easily lose weight. It's not just a matter of self-control and discipline.

> I couldn't believe it when I listened to [your program]—you are the first (and "only"—that I know of) person to really identify and explain in layman's terms what the heck is going on inside of my body. I knew this was the solution. I've been on your program for about two weeks, and have already dropped eleven pounds—feeling full and content, I might add. Thanks for the great program.
>
> —*S. G., Virginia Beach, VA*

The following letter illustrates how well our program works when male-centered diets have failed. This woman never mentions the words *relieving depression*, but clearly our program altered her emotional state, allowing her to feel good for the first time in years.

We are also very pleased that she put her fifteen-year-old daughter on the program with great results. This is the greatest gift a woman can give to a teenage daughter. It will lay the groundwork for her to have a healthy, positive relationship with food, her body, and her mind.

Here is Linda's letter:

Dear Philip,

I have been thinking of writing to you for nearly a year now. I can't put it off any longer as I feel guilty not letting you know what a *miracle* your [program] was for me.

I was not expecting much to happen. It was so shocking to hear your words on TV that all the things I had been dieting on were Bad Calories. I wasn't even eating fat, or fat loaded foods, just good old rice cakes, plain baked potatoes and carrots. And just as you said in the book [I was] gaining more and more weight than before . . .

The fun part is that I used to be a Weight Watchers [leader] for years. And as long as I weighed and measured everything *(yuck)* then I stayed trim. Then one day I decided not to be a Weight Watchers leader anymore. One day my fifteen-year-old daughter said to me, "Mom, I can't remember you being skinny." Ouch . . . Of course, I remembered it.

I also tried Diet Center, Nutri/System, you name it and I've done it. It was getting pretty hard to try anymore, as I could always lose weight but it always came back more and more.

I read your book . . . [and] thought that I would not tell a soul that once again I was going to lose weight. The Good Calories were . . . easy, fast and fit my lifestyle perfectly. In the summer I own and operate a resort on [a lake] in Montana. It is no sleep and no prepared food for three months. But I managed to live on Good Calories and the next thing I knew I had lost thirty-three pounds and really did not even have any pain involved.

No grouchies and mood swings . . . which I always had doing the Weight Watcher plan. Not to mention never being constipated as long as I ate Good Calories. It was great. Too easy to be true . . .

I have to tell you that my life was miserable as a fat person. I had prayed the week before you were on TV for help to get the weight off, and to keep it off forever. Now what I really want to know is if you were really on TV or did I entertain an angel?

I just had my fortieth birthday and have again a young body, and lots more energy . . .

I am so excited that I can eat eat eat and not feel guilty, and not weigh or measure anything . . . All of those greasy cravings are gone and your book and your hard work has given me back a life. I am so thankful to you and all your research . . .

My teenage daughter . . . also follows your [program]. She has lost fifteen pounds.

It is so easy that both of us are in total shock.

I can't wait to tell the next person about your book. Today in Yuma, Arizona, I met a lady from Texas, and she is going right out to buy one of your books.

I plan to tell the world before I am through. Thank you very much.

May God Bless you, Philip, for what you have done for me, my daughter, my friends, and many more.

CHAPTER 3

Serotonin, Food,
and a Modern Woman's Life

The Starvation Response afflicts all women, not just dieters. Today's women still carry the genetic instructions for the Starvation Response and low serotonin. This makes a woman's serotonin system respond differently to modern foods than a man's. The wrong foods more easily inhibit a woman's serotonin. Foods that nourish and support a man can actually attack a woman's serotonin and make her suffer.

Today's diet is very high in protein and high-sugar foods. Nearly every meal contains meat, fish, poultry, or dairy protein. Much of the modern diet consists of baked goods, candy, and processed foods, which are high in sugar. Unfortunately, these types of foods attack a woman's serotonin.

Excess protein evokes the Starvation Response simply because excess meat is one of the ancient signals that a woman faced starvation. Animal protein formed a higher percentage of our ancestors' meager daily caloric intake during times of starvation.

Our ancestors had a diet that consisted primarily of gathered plants and the occasional carrion, the remains of dead animals. It was hard to hunt with primitive weapons. Eating meat was a rare event. However, in a time of general starvation many animals died, and our ancestors feasted on the fallen bodies. At the same time the plant life disappeared. Therefore, excessive meat is a signal that brings on the Starvation Response.

Another signal for the Starvation Response is food high in sugar. Ancient fruits and vegetables had less sugar content than today's varieties. The only time our ancient ancestors ate fruits and vegetables with a high sugar content is when starvation forced them to eat rotting vegetables and dried remains of vegetables, both of which have higher than normal sugar content. Thus, high-sugar, insulin-producing fruits and vegetables are associated with times of starvation. In Chapter 5 we will see exactly what these high-sugar foods do to a woman's brain chemistry.

Many modern women eat diets high in protein and full of high-sugar foods. This means that even women who do not "watch" their weight can go through life impaired by low serotonin.

Most women will experience symptoms of low serotonin at some point during their lives, although they will not be clinically ill. Their symptoms will be milder—still strong enough to impair daily activities, but not strong enough to attract the attention of a doctor who might assume these problems are "just part of being a woman."

Low serotonin causes too many women to go through their lives experiencing a constant state of stress brought on by fighting negative impulses. We all know it is impossible to function well under stress. How can a woman exert all her true power if she is under constant stress?

How many women unknowingly suffer from minor depression, stress, and anxiety brought on by low serotonin? Although twice as many women are treated for depression as men, we're referring to a different form of depression—something that is undiagnosed and much more widespread than a condition requiring treatment. This is feeling a little "down" as a woman goes about her daily life. Instead of being fully in control of her mind and emotions, she finds herself being excessively sensitive and emotional and more prone to negative thoughts.

Since she isn't incapacitated or showing visible signs of illness, she assumes she is well. She is not. She is just unconsciously accepting a diminished experience of life. Gradually her capacity for joy and serenity erodes.

Although she does not meet the diagnostic criteria for clinical depression, this victim of the Starvation Response is suffering from a problem called "subsyndromal systematic depression." This disease is much more widespread, afflicting four times the number of people who suffer from clinical depression. People with this minor type of depression do not show all the common symptoms of clinical depression such as sleep disturbances, fatigue, and morbid thoughts. Instead, they show one or more of the symptoms of low serotonin.

OTHER PROBLEMS ASSOCIATED WITH LOW SEROTONIN

Many doctors think of food cravings and depression as two things that most often plague a woman's mind. Both of these afflictions are controlled by serotonin. Therefore, when most doctors think of women and serotonin they automatically limit themselves to these two problems. However, women suffer from the same problems as men, problems that encompass the whole variety of human existence.

Just having low serotonin can increase a woman's chances of suffering from one or more of the debilitating conditions associated with low serotonin. We will not bore you with how doctors made these determinations; for the most part, doctors took patients suffering from a wide variety of conditions and gave them serotonin-stimulating drugs. If stimulating serotonin reversed a condition, then it seemed logical that low serotonin might be associated with that disease.

Low serotonin creates many more subclinical conditions than just subsyndromal depression. Most women experience mild cases of only a few of those symptoms, but almost all women suffer from some minor problem. Random fate and preexisting conditions determine each woman's experience.

Impaired Social Interactions

Serotonin mechanisms appear to help regulate social dominance. Increasing serotonin activity in male monkeys can make them become the dominant males when placed in a new social group.

Serotonin has also been linked to the formation of social groups, grooming, maintaining social proximity, approaching others, and other forms of social cooperative behavior.

Appropriate Behavior

Serotonin helps us avoid unpleasant situations by controlling our memories of things to be avoided.

Excessive Aggression

It appears that serotonin is necessary to stop impulsive behaviors, particularly aggressive and violent behavior. Low levels of serotonin are linked to excessive aggression and other destructive behaviors. Low levels of serotonin are associated with impulsive murders, but not with murders that were planned out. Decreased serotonin activity is also found in excessively aggressive mentally retarded patients. Arsonists have lower levels of serotonin.

Are men more aggressive than women because they have lower levels of serotonin?

Menopause

During menopause the hormones that elevate serotonin decrease, which tends to impair serotonin synthesis. Some doctors feel that many of the psychological symptoms of menopause are nothing but a reflection of these changes in serotonin. It could also explain why women gain weight during menopause. Many doctors use Prozac and similar serotonin-stimulating drugs to treat the symptoms of menopause.

Suicide

If you suffer from low serotonin you appear to have an increased risk of committing suicide, the ultimate form of self-directed aggression. The brains of suicide victims show low levels of serotonin. Figures cited in the April 18, 1994, issue of *Newsweek* reveal that approximately 20 percent of depressives hospitalized with low serotonin will commit suicide within one year, but only 1 to 2 percent of depressives hospitalized with normal serotonin will commit suicide.

Although not all people who attempt suicide suffer from low serotonin, there are significant differences between suicide attempts with low or normal serotonin. When you suffer from low serotonin you tend to plan your suicide long in advance and are more likely to repeat the attempt if you were unsuccessful the first time.

Some researchers claim that contemplating suicide is enough to elevate serotonin levels. If true, this would mean that people who systematically plan their suicide do so because it brings some relief from their depression.

Men are two to three times as likely to commit suicide as women, perhaps because men are born with less serotonin than women.

Obsessive-Compulsive Disorder

Approximately 2 to 3 percent of people suffer from obsessive-compulsive disorder (OCD); they exhibit an abnormally high degree of fear, doubt, anxiety, and a need to control their environment. Common obsessions include sexual obsession, aggressive behavior, excessive fear of failure, and the need for sympathy. Common compulsions include excessive checking on previously performed tasks, excessive washing, constant counting, the need to ask questions or confess, hoarding, and the constant rearrangement of articles in their environment. OCD patients respond well to serotonin enhancers such as Prozac, indicating a role of serotonin in this disease.

Overeating After Stopping Smoking

A person who stops smoking may suffer from symptoms, especially carbohydrate cravings, which are consistent with low serotonin levels.[1] Elevating serotonin prevents the weight gain normally associated with tobacco withdrawal.

Sexuality

Female sexual function relies on the perceptions of reward and pleasure mediated by serotonin. Whereas high serotonin inhibits a man's sexual drive, it stimulates a woman's sexual responsiveness.[2] Low serotonin will make it difficult for women to be sexually receptive.

Sleep and Melatonin

Sleep disturbance is linked to low serotonin. Serotonin is made into another neurotransmitter, melatonin, that controls our internal clocks and determines when we awaken and when we feel sleepy. It is entirely possible for serotonin levels to fall so low that melatonin levels also fall.

Melatonin is also thought to be responsible for a wide variety of processes ranging from aging to cancers.

Premenstrual Stress

Serotonin levels rise and fall during the menstrual cycle, reaching their lowest point just before the start of menstruation, at the time when PMS peaks. This suggests that PMS is a condition of low serotonin. Supporting this idea are the observations that: (1) low serotonin creates cravings for sweets similar to that experienced by many PMS sufferers, and (2) Prozac and other serotonin-enhancing drugs are effective treatments for PMS.[3]

Many doctors suggest that PMS is not an inherently predetermined part of being a woman, but is rather a cultural phenomenon. Some women find this to be an offensive suggestion because it seems to imply that PMS is a result of cultural hysteria, not a physical affliction. Doctors respond to this criticism by pointing out how widely the incidence of PMS varies according to culture. For example, Amish American women have much lower incidence of PMS than American women who live nearby. Does this reflect the differences between these cultures or the different foods they eat? We suggest that many of today's women suffer from a continual case of PMS brought on by modern dietary habits.

Other Conditions

The following is a list of other medical conditions that have responded to treatment by increasing serotonin levels, suggesting that low serotonin contributes to their symptoms.

- arthritis
- bipolar disorder (cycling between mania and depression)
- borderline personality disorder (unstable mood and behavior)
- cataplexy (extreme muscle weakness attacks)
- chronic pain
- depersonalization disorder (detachment from reality)
- dysthymia (chronic low-grade depression)
- elective mutism (refusal to speak)
- emotional liability (sudden changes in emotion)
- hyperactivity
- hypochondria
- hypomania (abnormal mood elevation)
- panic disorder
- posttraumatic stress syndrome
- schizophrenia
- seasonal affective disorder (winter depression)
- self-injurious behavior
- social phobia

How can a woman experience a sense of well-being when she is prone to this long list of pathologies?

Living with low levels of serotonin is like operating with a low-grade flu. A woman can function, but in no way is she as functional as she should be. In no way is her experience of life as joyous and full of energy as when she is totally well.

We are not suggesting that the wrong foods or weight-loss diets create a psychiatric crisis. How poorly a woman feels depends on her preexisting levels of serotonin. The wrong foods can create a major problem only when a woman is already suffering from low serotonin. Then two small depressions can join to create a crisis.

Think of it this way. Dieting and eating the wrong foods lower serotonin and remove the barriers designed to prevent small mental problems from escalating. It is the same as removing the police from the streets of a troubled city. Without police protection a minor disturbance escalates into a riot.

Does this mean half the population has a mental disease? No! All it means is that women's brain chemistry can become temporarily so abnormal that they cannot perform to their full capacity.

LOW SEROTONIN AND FEMALE STATUS

Modern societies discriminate against women, excluding them from positions of real authority. Women were not always held in such low regard. In many ancient cultures, women were either socially dominant or equal with men. Ancient humans worshiped women as goddesses: Shakti, Isis, Aphrodite, and the Corn Mother. Even in this century, anthropologists can point to many "primitive" cultures where women are treated more equally.

What has changed? One of the changes is the food we eat.

When women cannot perform at full capacity it limits their status in society and removes them from social and political power. The link between food and power is so strong that throughout history women lost status whenever they ate foods inhibiting serotonin. Conversely, women also gained status when they ate enough of the proper serotonin-enhancing foods.

We believe the food consumed by a society plays a pivotal role in its development. Modern food brings on low serotonin levels and impairs women so they do not have the same emotional and mental energy that they did in those societies where they shared power with men. Modern foods and weight-loss diets make women become more emotional, more depressed, more fatigued, and more prone to extreme behaviors. At the same time, men suffer from none of these problems. Thus, men retain an unnatural competitive advantage over women. This may be one of the most important issues facing today's women.

Recognizing that *all* women, not just dieters, need the right foods is the big advance this book offers.

Yet, the awareness of the relationship between female behavior and food still remains at the level of myth and subconscious thought. Most women think of weight control when they think of food. Very few women consider food to be a powerful drug that can alter every aspect of their lives.

Is it purely coincidental that whenever women from different eras and cultures ate woman-friendly foods they were more likely to be treated as equals of men? We think not.

FOOD, STARVATION, AND FEMALE POWER

We first became aware how strongly food altered a woman's power when we recompiled data originally gathered by Peggy Reeves Sanday in her book *Female Power & Male Dominance*.[4] Sanday examined the anthropological literature for more than 150 "primitive" cultures and repeatedly asked the same simple question: What conditions must exist within a culture for manifestations of female power to be present? Although Sanday did not consider whether food was the determining factor, it is possible to take her data and recalculate it to illustrate the role of food in determining a woman's place in society.

Starvation dieting is the easiest way to bring on low serotonin. Therefore, we would expect women not to be as powerful in cultures that periodically starve. Sanday's data shows a correlation between starvation and decreased female power. A society with a steady supply of food is almost 50 percent more likely to give women political power than a society that undergoes periodic starvation (Figure 3-1).

To see what different foods do to a woman's status we divided the anthropological data into six different categories of eating habits. Table 3-1 gives the definition of each eating pattern.

As societies develop more sophisticated forms of agriculture, they tend to favor crops with a higher sugar content. They also tend to use these crops to feed livestock, increasing the amount of meat in their diet.

Therefore, societies tend to move from woman-friendly to woman-unfriendly diets with large amounts of animal protein and higher-sugar fruits and vegetables.

Our theory also predicts that as a society moves to woman-unfriendly foods, women's status will decline. The data proves this is true. Figure 3-2 shows that women have the most political power in a gathering society where the principal foods are crude fruits, roots, and vegetables. When societies start farming and eating vegetables that are higher in sugar, women's power also declines. However, the greatest decline comes in those societies that hunt or raise herds of livestock. These cultures eat large amounts of meat.

Women's status can be reflected in religious as well as political

FIGURE 3-1

Percentage of Societies with Females Holding Significant Political Power

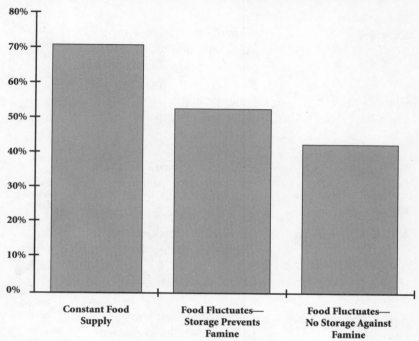

power. Women are part of their religious mythology when the society respects women. Therefore, Sanday examined what types of societies included or excluded women from their creation myths.

The results were clear. Women have more status in societies that eat woman-friendly diets. No meat-centered hunting societies worship women in this way (Figure 3-3). Gathering societies with woman-friendly foods are more than three times more likely to worship women than pastoral herding societies that rely on meat as nourishment.

These figures show there is a correlation between food and female power. It does not prove our theory; unfortunately, nothing can. Scientific proof would require creating isolated groups and raising women on different foods. Clearly, this is impossible.

However, we wonder whether the same processes are changing the status of modern women. Protein intake has risen enormously. At the same time the amount of woman-friendly carbohydrates in the

TABLE 3-1
Diets of Different Societies

Type of Society	Diet	
Gathering	primarily vegetarian—low-sugar fruits, nuts, and roots; little protein; carrion sources of meat	**Woman-Friendly Foods**
Semi-intensive Agriculture	some planting of low-sugar plants; still principally vegetarian diet	
Cultivation of Fields	self-sufficient farms with some livestock; more meat in diet; more sugar in foods	**Less Woman-Friendly Foods**
Advanced Agriculture	still more livestock and meat; still more sugar in plants	
Hunting	high-protein diet	**Woman-Unfriendly Foods**
Animal Husbandry/ Pastoral	high-protein diet; fewer vegetables in diet due to constant movements of camp	

American diet has been steadily declining. When compared to only a century ago, today's women now eat less than half of the woman-friendly foods they once ate (Figure 3-4). When compared with the diets our ancestors evolved to eat, today's women are eating less than one third of the woman-friendly carbohydrates they need. Perhaps this is why more of today's women suffer from depression, and why women have yet to achieve social equality.

MEDICINE IGNORES WOMEN'S UNIQUE NEEDS

We realize that attributing the source of a woman's well-being to nutritional rather than psychological causes is a radical notion. It should not be. The only reason this idea seems unusual is that modern medicine has long ignored the real biochemical differences between women and men.

Because modern foods and weight-loss diets do nothing negative to a man's serotonin, modern medicine assumes nothing is wrong with the food women are eating. Therefore, most doctors are totally unaware of how serotonin imposes unique nutritional requirements on women.

When the medical profession looks at the symptoms of low serotonin—depression, PMS, emotionality, and so forth—doctors often pass them off as "women's complaints." This leads to constant misdiagnosis. Because the Starvation Response's low serotonin afflicts only a woman's brain, doctors often assume that women are more emotional than men. This discrimination is so bad that in December of 1990, The American Medical Association's Council on Ethical and Judicial Affairs said:

FIGURE 3-2

Percentage of Societies with Females Holding Significant Political Power

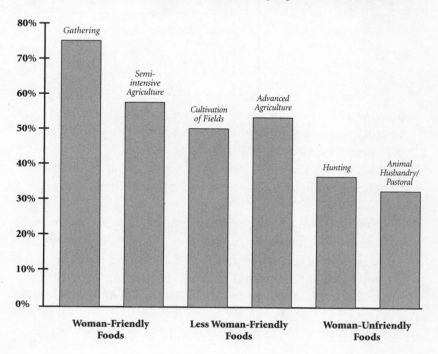

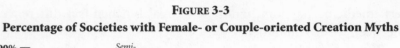

FIGURE 3-3

Percentage of Societies with Female- or Couple-oriented Creation Myths

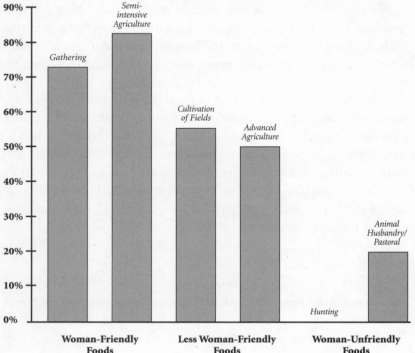

Gender bias may not necessarily manifest itself as overt dis-crimination based on sex. Rather, social attitudes, including stereotypes, and other evaluations based on gender roles may play themselves out in a variety of subtle ways.

For instance, there is evidence that physicians are more likely to perceive women's maladies . . . as the result of emotionality.

This view of women as being excessively emotional makes it almost impossible for women to get adequate care. Numerous stud-ies conducted over the last twenty years have shown that when men and women present themselves to doctors with the same physical or emotional complaints, women are significantly more likely than men to receive prescriptions for antidepressants, tranquilizers, and other psychotropic drugs, suggesting a pervasive physician bias in diagnosing and prescribing. Women receive 73 percent of all prescriptions written for psychotropic medication—and an incredible 90 percent when the prescribing physician is a psychiatrist.[5]

These doctors ignore women's real needs because medical research, on which they base their diagnoses, also ignores women. As recently as 1985, the Public Health Service of the United States reported that lack of research data on women limited our understanding of women's health problems. In June 1990, the General Accounting Office, Congress's investigative arm, found that The National Institutes of Health systematically excluded women from their studies.

The National Institute on Aging's Baltimore Study excluded women for more than twenty-five years because their building had only one toilet. The Physicians' Health Study promoting aspirin to prevent heart disease included 22,000 men, but no women. The Harvard study on a possible link between caffeine and heart disease included 45,000 men and no women.

Many of our current notions about diet are based on the 1982 Multiple Risk Factor study that linked cholesterol and heart disease, suggesting that LDL cholesterol (often called "bad" cholesterol) leads to heart attacks. Unfortunately, this study included 13,000

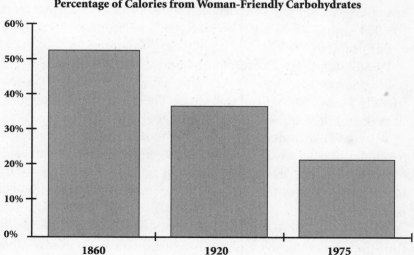

FIGURE 3-4
Percentage of Calories from Woman-Friendly Carbohydrates

men and no women. Not surprisingly, there is controversy about whether its results apply to women. Many doctors feel that there is no association between high blood cholesterol and cardiovascular deaths in women, and that it is not wise to screen for and treat high blood cholesterol in women. Yet, doctors continue to screen women for high LDL cholesterol and recommend that women follow diets designed to lower LDL cholesterol. Why? Because doctors are trained to treat the male body, not the female.

If scientists can't agree on whether cholesterol causes female heart disease, how do we know their other dietary recommendations apply to women as well as men?

Why is there such a prejudice against studying a woman's medical needs? Doctors are afraid a woman's hormonal cycle may confuse the results. Of course, it may also show what is relevant to women. Far too many drugs behave differently as a woman's hormones change. Wouldn't it be appropriate to know what treatments work for women under the conditions in which women will use them?

THE ULTIMATE PROOF OF HOW FOOD CONTROLS A WOMAN'S WELL-BEING

As far as we are concerned, the ultimate proof that food is a significant factor in controlling a woman's place in the world, her sense of self-esteem and self-confidence, is the example of the more than 100,000 women who have tried an earlier and less powerful version of our program.

We still receive letters from these women, and although they were only trying to lose weight, they also tell us how eating woman-friendly foods has changed their lives. Their experience taught us the value of proper foods and constitutes all the proof we need.

We would like to share with you the following letter from a Ph.D. psychologist who was amazed to discover how our program altered her state of mind. She lost weight—ten pounds in a month—but hardly noticed because the change in her emotions was even more dramatic.

We value her letter because she is trained to look at changes in the mind and emotions, and is somewhat dubious about men who have something to say about a woman's weight. As a published researcher in psychology, she previously felt the emotional problems dieters suffered were purely psychological and not physiological. She no longer feels that way:

> I was curious and just wanted to see if anything would happen. I figured the proof would be in how your diet plan worked on my body. A body that had never succeeded on diets.
>
> The first thing I noticed was that I had small periods in the day when I wasn't worried or anxious. I wasn't having the horrible feeling that something was going to fall apart at any second. I noticed that my depression started to lift, and I happened to notice without really trying I was indeed losing weight, ten pounds in fact.
>
> Rather than feeling alienated from my body, I began to feel as if I was in my body, my body was beginning to reflect me. My body started to become a friend instead of an enemy that was condemning me to the stereotypes of old age.
>
> I also began to experience energy. Along with this energy, I felt a whole repertoire of feelings, including feelings of love towards friends and family. I started to spend more time with friends. I began to be able to receive and accept the kindness, goodwill, and love that others had towards me. Normally an introverted, overly analytical, slightly depressed person, I was shocked to experience feelings of jubilance and happiness.

CHAPTER 4

A Call to Freedom

As we have said, serotonin controls the way women eat, drink, think, and seek pleasure.[1] Since these behaviors encompass much of a woman's life, she cannot function optimally when her serotonin is low.

The power of serotonin to regulate so many aspects of a woman's life has the drug industry salivating at the prospect of selling serotonin-boosting drugs for every one of these afflictions. Propelled by millions of dollars in drug company PR, magazines and their readers jumped on a bandwagon that proclaimed serotonin pills to be a miracle. And millions of women unwittingly became part of a great experimental step into the unknown—for no one really knew what these drugs would do over the long-term.

Now we know the results of that experiment. Serotonin-stimulating diet drugs have been withdrawn due to dangerous side-effects. And we're proud to say that Philip had a small part in correcting the wrong that these drugs did to so many women. First, the original edition of this book, which was published before the dangers of these drugs were widely known, proclaimed that these drugs were potentially dangerous. Secondly, Philip published an article in *The Journal of the American Medical Association* suggesting that these drugs create the need for higher levels of serotonin synthesis than dieting women can ever produce. Dieting women, the very population most likely to take diet drugs, suffer the most because they cannot replace the serotonin used by the serotonin-stimulating diet drugs.

THE DANGERS OF SEROTONIN DRUGS

Unfortunately, we still do not think that the potential danger of these drugs has been fully explored. These drugs were withdrawn due to suggestions of increased risk of valvular heart disease. Some doctors raised the question of neurotoxicity in animals, while other doctors and the media ignored the real issue, what happens to the brains of people who take these drugs? Therefore, we would like to once again point out the potential risk of these drugs in case there is a movement to reintroduce them to the marketplace.

Equally serious problems plague all the serotonin-boosting drugs simply because they create an unnatural state of serotonin regulation. There are some preliminary indications that after several years of taking these drugs your body may adapt and change how you use serotonin when not taking these drugs. We wonder whether this could, over years, create an addiction-like dependence on these drugs.

In the fifties, doctors thought amphetamine diet pills were harmless; now we know they are addictive. Methadone, an addictive narcotic now traded on the streets, was originally developed as a safe alternative to heroin. Heroin was originally developed as an alternative to morphine. Morphine was originally developed as an alternative to opium. In every case, the danger of these drugs was known only after millions of patients had been taking them for many years.

The Food and Drug Administration was so concerned about the side effects of serotonin diet drugs that their advisory panel initially recommended against approval of Redux. They felt it was too dangerous, creating brain damage and potentially fatal pulmonary hypertension. The only way it got approval was through a vote that was rescheduled when its opponents were away at a scientific conference. Since the introduction of Redux, doctors have found that its side effects are even worse than feared, and the Food and Drug Administration has required that Redux carry a special warning.

Another serotonin diet drug, sibutramine, was not approved for sale due to concern about its side effects. Yet sibutramine was easy to monitor for side effects, according to Lynn McAfee of the Council on Size and Weight Discrimination, who wanted "an alternative to Redux."

Even Prozac has problems as a diet drug. Although the normal dosage of Prozac is about 20 milligrams per day, using Prozac for weight loss requires a dosage of 60 milligrams per day. Originally, Prozac's manufacturer sought to create a new diet product by repackaging Prozac in a higher dose and renaming it. However, the application for approval was withdrawn after concerns surfaced over the side effects of the increased dosage.

We are afraid that high doses of serotonin-stimulating drugs might bring on as many problems as did the amphetamine diet drugs of the 1950s. At that time "speed" and other amphetamines were thought to be a safe way to produce miraculous weight loss. These drugs were considered so safe that even President Kennedy used them during his presidency. However, it turned out these drugs were highly addictive. They are now banned but remain a staple of the illegal drug trade.

There is some preliminary evidence that the same problems with addiction may arise with serotonin diet drugs. Redux and fenfluramine create significant brain damage in laboratory mice and rats. After the drugs are withdrawn, serotonin levels fall and stay low for weeks. This creates the need to take more serotonin-stimulating drugs, making the subject become addicted.

Another way that serotonin diet drugs could create addiction is by destroying the cells that create serotonin. Dr. Mark Molliver of Johns Hopkins University found that a small overdose of Redux could kill serotonin-producing brain cells. The required overdose is so small that just taking two pills instead of one could create damage, perhaps attacking the brain in the same way as the illegal drug Ecstasy, creating the equivalent of a permanent state of serotonin depletion. The only way to treat this brain damage might be to continue on serotonin-stimulating pills for life. In other words, the dieter would become addicted to the use of serotonin stimulants. This potential damage is why the full Food and Drug Administration advisory panel voted against approving Redux.

Even when Redux was finally approved, it was intended to be sold *only* to the morbidly obese who are in danger of death due to stroke or heart attack. In such a case, the danger of taking Redux is outweighed by the danger of morbid obesity. However, very few people taking Redux are morbidly obese. Most are people who want to lose a few pounds, not hundreds of pounds. These women should not be exposed to the risk of Redux or the other serotonin diet drugs.

SEROTONIN DIET DRUGS WOULD STILL REQUIRE WOMAN-FRIENDLY FOODS

Even if doctors invent a new "safe" drug, dieting women will still need to eat woman-friendly carbohydrates. Any drug can never create serotonin; only tryptophan can do that. These drugs merely make serotonin work harder. The less tryptophan there is, the less serotonin there will be for these drugs to alter. They do not work properly unless there is enough tryptophan and serotonin. In fact, several studies have shown that Prozac-like drugs do not work as efficiently in women who eat the wrong diets.[2]

Serotonin-enhancing drugs are powerless with a diet that lowers serotonin. These drugs work by making serotonin work more effectively; if there is no serotonin there is nothing for these drugs to work on. This means that women must eat the proper foods if these drugs are to work.

The power of the wrong foods to negate serotonin drugs is best illustrated by the experience of twenty-four mental patients at Yale University who were fed a diet that did not support serotonin synthesis.[3] These patients had been severely depressed but recovered when they started taking serotonin-stimulating antidepressive medications. When they ate the wrong foods, sixteen out of twenty-four patients (66 percent) immediately sank back into depression as their serotonin levels fell. It did not matter that they continued to take their serotonin-boosting antidepressants. These drugs were powerless without the foods needed to support serotonin synthesis.

Can you imagine how frustrating it was for these patients to fight their way back from depression only to have all their success fly out the window after just one day of eating the wrong foods?

What will happen to dieting women who eat the wrong foods

while taking serotonin-stimulating diet drugs? The Yale study also showed eating the wrong foods blocked serotonin-stimulating drugs more easily in women than in men.

What will happen when weight loss decreases a woman's serotonin? Dr. Philip Cowen's group at Oxford decided to study serotonin synthesis in women taking serotonin diet drugs.[4] They gave the study's participants Pondimin (fenfluramine) and examined the effects of three weeks on a 1,000-calorie-a-day diet. They found that women, but not men, experienced a significant decrease in the tryptophan needed for serotonin synthesis. Women, but not men, also experienced changes in the hormones controlled by serotonin, and these changes were consistent with a decrease in serotonin. Perhaps the reason serotonin diet drugs stop working after a few months is that serotonin levels have fallen too low during weight loss.

As this book went to press, a group of Spanish researchers published a paper that showed that women taking Redux suffered from a decrease in tryptophan's ability to enter the brain after only three months of dieting.[5] After six months there was twice the decrease. What would have happened after one year? Even if there were no further decreases, this gradual decline could impair serotonin synthesis and interfere with Redux's ability to work. Perhaps this explains why the rate of weight loss gradually declines after three months of Redux treatment.[6]

The difference between men's and women's response to these drugs is so great that researchers in Great Britain, after finding that women and men have different responses to Pondimin, suggested that "the findings cast doubt on results already obtained employing this dose of the drug in studies that have included male subjects."[7] This means that serotonin drugs have been approved, in part, by examining studies on male subjects, and these studies may not be relevant to women, especially dieting women. Let's hope that any future drug is tested on women.

This research group goes on to say, "This introduces food or fasting as a compounding variable [on each sex's response to serotonin-stimulating drugs]." They conclude by calling for more research on the effects of food, fasting, and the time window in which these things can have an effect on serotonin-boosting drugs.

In simpler language, these researchers confirmed our suggestion that the elements necessary to create a woman-friendly nutritional environment may also be necessary for women who take serotonin-stimulating drugs. Even if a woman takes serotonin-elevating drugs such as Prozac, fenfluramine, or Redux, she must eat the proper foods to experience their full benefits.

TRYPTOPHAN PILLS

Are there any pills that might work? Tryptophan is the amino acid that is made into serotonin. Tryptophan pills increase the amount of tryptophan entering the brain and thereby increase the production of serotonin. Tryptophan pills were the original way doctors treated low serotonin.

If diets decrease tryptophan entering the brain, why don't dieting women just take tryptophan pills? In the past, many people did just that. Not only did they lose more weight but they also felt relief from the other symptoms of low tryptophan, especially depression.

We cannot recommend tryptophan for one simple reason. In the 1980s the Food and Drug Administration banned tryptophan because several dozen people died after taking it. Since that time, numerous reports have suggested that tryptophan is not inherently dangerous; these deaths resulted from contamination. However, any discussion of whether tryptophan is safe or unsafe is useless, because tryptophan pills are illegal. Instead, we recommend woman-friendly foods.

5-HYDROXYTRYPTOPHAN

5-hydroxytryptophan is what tryptophan is made into before it becomes serotonin. When you take 5-hydroxytryptophan it acts to stimulate serotonin. It also acts to stimulate another neurotransmitter system, the dopamine system. The combination of these serotonin and dopamine effects makes 5-hydroxytryptophan into a potent antidepressive.

Obese people have less 5-hydroxytryptophan than do normal-weight people, something to be expected because the obese have diminished serotonin synthesis.[8] Obese humans who take very high doses of 5-hydroxytryptophan spontaneously lose weight, probably due to decreased appetite.[9] However, at these dosage levels up to 40 percent of subjects experienced at least one episode of nausea or other side effect. Unfortunately, we are not aware of any weight loss studies at lesser doses. In many ways, the reported side effects were similar to those experienced with Prozac, Redux, and other serotonin-uptake inhibitors.

5-hydroxytryptophan's legal status is unresolved. It is different from tryptophan and so was not covered by the tryptophan ban. The FDA told us they were unclear if 5-hydroxytryptophan was legal to sell. However, we are aware that some manufacturers are thinking of introducing 5-hydroxytryptophan products; check with the Resources in Appendix III. We are not suggesting that 5-hydroxytryptophan is better than a purely nutritional approach. Such pills work best when you eat right. Furthermore, the fewer strong chemicals you introduce into your body the better off you are.

Additionally, there is evidence 5-hydroxytryptophan can be affected by the same contamination problems as tryptophan. There are also several studies showing that 5-hydroxytryptophan interacts with stimulants and psychoactive and psychiatric drugs to produce delirium and other significant psychiatric problems requiring hospitalization.

FUTURE DIET DRUGS ARE NOT THE ANSWER

Almost all of today's diet drugs work by altering serotonin, but this may change in the future. A wide variety of drugs are under development (Table 4-1). At first glance these drugs seem like the answer to a woman's prayers. However, they may create problems simply because they are effective. It appears that *any* weight loss can create low serotonin in women. Unless they eat the right foods or take serotonin-stimulating drugs, women will suffer with these drugs. They are not the answer.

TABLE 4-1
Diet Drugs Under Development

Drug	Mode of Action
Orilstat	inhibits absorption of fat
Leptin	a hormone that decreases appetite and boosts metabolism
OB-Receptor	makes leptin work better
BTA-243	stimulates fat burning
CCK agonists	mimic hormones that inhibit appetite
Butabindide	helps CCK to inhibit appetite
NGD-95-1	alters brain's perception of eating
Neuropeptide Y inhibitors	prevents appetite and burns fat
Troglitazone	burns fat

PRESERVING THE FEMALE FOOD REVOLUTION

We are in the middle of a women's revolt against the failure of medicine to produce lasting weight loss in women. No longer do women assume that each fad diet is the answer. There is a healthy skepticism about whether diet programs do anything but make diet food suppliers rich. Drug companies are riding the crest of this revolution, trying to turn discontentment about food into an opportunity to sell drugs.

We are afraid that this anger will cause women to lose track of the larger issue, the necessity to create a woman-friendly nutritional environment.

Unless women change their eating habits they will have to operate with a handicap. If they take drugs they will be subservient to drug companies. Women can and should be able to assume control of their own destiny and diet and not be dependent on pharmaceuticals.

Women are not so flawed that they require continual medication. All that is flawed is our understanding of what it takes to support a woman's unique body chemistry.

This is not the first time that a women's revolt against male-oriented medicine has been turned into a commercial opportunity. The first major women's revolt started when Ester Manz of Milwaukee thought it was time to organize a program similar to Alcoholics Anonymous. She felt that if women of the 1950s gathered together and talked with each other and exchanged support and ideas, it would be easier to lose weight if they were independent of the diet industry. So she organized Take Off Pounds Sensibly (TOPS), where women gathered to pool information and design their own diet programs. TOPS clubs sprang up around the country to free women from fad diets.

More than 60,000 members gathered together and held trials at "Courts of Weights and Measures." Women who lost weight were congratulated. Women who regained weight were gently abused. TOPS was nonprofit and promoted no diet plan. Everyone lost weight under the supervision of her own doctor. In many ways this reflected the growing belief that there was no miracle solution for female weight loss, so it didn't really matter what plan a woman used.

Although TOPS was a sincere attempt of women to exert power, it also proved there was money to be made by channeling female frustration into captive groups of diet consumers. The mutual support aspect of a diet group could also be used to create peer pressure to force women to buy a diet product.

In 1963, Jean Nidetch, a Brooklyn-born housewife, founded a support group known as Weight Watchers. It differed from TOPS in that it was profit making and promoted the same diet for all women. Women were encouraged to buy frozen foods and cookbooks. Weight Watchers proved to be so profitable that a food conglomerate, the H. J. Heinz Company, bought it. Later Heinz would also attempt to buy Nutri/Systems, another diet club.

Recently, the Federal Trade Commission attacked the entire diet club industry for promoting false advertising. Their complaint said it did not matter whether the club was Jenny Craig or Weight Watchers; the vast majority of diets failed.

Now women have turned away from these diet clubs and embraced diet drugs. Are we going to repeat history and see yet another women's revolt fail as big pharmaceutical companies jump on this marketing opportunity?

Please do not let this happen to the woman-friendly foods revolution.

CONCLUSION

Women should understand it is not their fault previous diets didn't work. Their failure does not reflect a lack of willpower or some inherent weakness in them. It merely indicates they are eating the wrong foods and their brains have too little serotonin.

Women should be happy to learn their gender's unique brain biochemistry allows them to elevate serotonin with food. Men cannot do this as easily.

When women take advantage of the proper foods they will not only look better but will be able to live their lives free of the brain imbalances that previously kept them from expressing their full potential.

We constantly hear from women who can lose weight only with the proper foods—women-centered foods that support serotonin without creating the excess insulin of food addiction.

> For the past six years of anguish, I had been to the "so-called best" endocrinologists and nutritionists in NYC. A notable Ph.D. in nutrition advised me that I was lying to her about what I ate, and I have a "bad problem of denial." Needless to say, this has left me upset and seeking more drastic measures to shed this dreaded thirty lbs. I have tried thyroid medication, low-calorie diets, and liquid fasts.
>
> Just by [following your program], I feel validated for years of screaming about the low-calorie diets and fasting I was on. I will tell you I had been to very reputable clinics and doctors who made me feel weak. Never once did the doctors look at the effects that chronic dieting had on me. Thank you so much.
>
> —*P. G., Altamonte Springs, FL*

Doctors and pharmaceutical firms do not like to hear women complaining that conventional programs do not work. They absolutely do not like to hear that someone may offer a solution that deprives them of their stranglehold on women's bodies. We realize that people who question the pharmaceutical and diet industries are opening themselves up to a well-financed attack by these powerful interests. However, we feel we are at a crossroads where the growing recognition of serotonin can either enslave women to drug companies or free them.

A woman will be enslaved if she commits herself to a lifetime of drugs or harmful dieting. It is an issue that affects the quality of every woman's life. It is one of the great issues of our day. Humankind will grow in freedom and equality when all women know how to feed their serotonin.

Women will be freed when they eat foods that allow them to have slim bodies and powerful minds.

Our fondest hope is that someday all women, not just dieters, will realize that they must eat to maintain normal serotonin levels and to keep their bodies and minds functioning fully. Only then will women be physically and mentally able to assume their proper place. Only when that happens will the world fall back into the natural order determined by both the male and female energies.

Monika discovered the power of eating the proper foods long before she started to date, and eventually marry, Philip. This is her story.

A number of years ago, I found myself confined to a wheelchair recovering from a serious injury. After some time I discovered, to my great distress, that I had another problem on my hands. I was gaining weight rapidly. I had ballooned up to 150-plus pounds and, given the months of physical confinement ahead, I knew I was in trouble. Until the accident I had been fanatical about exercise. It was the only way I could keep from gaining weight. Even with all that exercise, though, I still hadn't been able to get rid of those extra fifteen pounds. No amount of dieting or exercise ever got rid of that.

Of all people I should have been able to. I was president of a company that sold health-care supplements and had kept up on the latest developments in healing, especially food and supplements. In an effort to lose weight I had tried all the latest "natural" methods for weight loss: metabolic herbs, raw foods, juice fasting, colon cleansing, seaweed wraps. Although these things have other benefits, nothing seemed to make any difference in my weight. I would lose a few pounds and regain them back in a short time.

And, like almost every other woman I know, I had tried the basic caloric restriction approach. I tried. And I tried. It would work for a few days, but then I just couldn't function properly. My mind wasn't as clear and focused. My energy level would drop and I'd get depressed. I wouldn't suffer too long, however, because the intense sugar cravings would drive me to eat the fattening foods I was avoiding. Chocolate and cookies brought me out of the mental haziness and my latest attempt at dieting. I would always be left, however, with a lingering feeling that I didn't have enough willpower and strength to see it through. This was especially bothersome because I consider myself to be a disciplined person. What I didn't know was that discipline alone wasn't going to make me lose weight.

So when I hobbled onto the scale after the accident I knew I was in serious trouble. That's when I decided to call Philip.

At the time Philip and I were just good friends. I'd always admired his innovative ideas on many subjects. He's a brilliant scientist whose acclaimed work in the area of DNA and aging was far ahead of its time. Although trained as a biophysicist, he had been researching weight loss. True to form, his ideas on weight loss were very different from anything else I'd heard. I had to admit, though, that his diet had worked for him; he had lost thirty-five pounds and looked great.

So I began the program. The hardest thing for me to do was to force myself to eat breakfast and never to skip a meal. I kept telling myself that I couldn't let my body feel it was starving, that it needed to be fed regularly, so I actually ate more than before.

After a few days I noticed my cravings for sweets died. This was exactly the opposite of what other diets did! With the other diets I'd tried, the cravings would begin after a few days, and I would be constantly hungry and begin to obsess about food. Now I noticed the cravings were gone and, after some time, my appetite got smaller. It wasn't an act of self-discipline; it was my body's own intelligence responding. It was the first sign that my body had become more balanced.

Exactly one week later, I hobbled back on the scale and was absolutely thrilled to discover I had lost two pounds! This might not seem like much, but to someone like me, whose body seems to hold on to extra weight like a miser to his money, this was a revelation. After all, I was now confined to a wheelchair and eating more food than when I was exercising for an hour and a half a day.

I continued to eat according to his program during the rest of my recuperation, which wasn't difficult to do. I ate healthy, satisfying meals and never went hungry. To my delight, after six weeks confined to a wheelchair, I lost twelve pounds and continued to lose a total of twenty pounds!

I also loved the effect this diet had on my mind. Previously there was so much negative energy associated with food and with my failure to lose weight. My mind had been filled with a constant stream of negative thoughts about food for as long as I could remember. I never seemed to have any peace around the issue of food and weight. There was tremendous fear and anxiety surrounding it. Now, it was different. Each day I watched my thought patterns change. I began to relax around the issue of food. And I finally felt this unattainable thing called "losing weight" was pretty simple. I now could enjoy eating like a thin person eats, without all the internal craziness. I felt more balance; I was happy and content. I felt liberated from the madness around food and weight loss!

I found myself being happier as I went through my daily routine. And this happiness was more than joy at my weight loss. Something had changed within my mind. I felt full of joy and power. More things seemed possible. Problems seemed to be smaller than before and I looked forward to every day, knowing that I was in control of my destiny.

Occasionally I am forced to eat the wrong foods during the holidays or because I'm traveling. Then I notice the cravings return, I've put on a few pounds, and feel less in control of my food. So I just go back to eating lots of the woman-friendly foods included in the programs and, after the second or third day, the cravings are gone, my appetite is decreased, and I feel more in control.

Sometimes I think these foods are magic because of the incredible effect they have on a woman's mind and body. I became involved in the book because no previous book explained how women need special food to sustain both their bodies and their minds. It's very satisfying for me now to let other women know about this simple secret, so they can be liberated from this madness as well.

Part II

THE PROGRAMS

CHAPTER 5

Creating a Woman-Friendly Nutritional Environment

The ideal diet would retain the insulin-fighting power of high-protein diets while adding the ability to stimulate enough serotonin synthesis to prevent serotonin levels from dropping. Such a diet would avoid the problems of both high-protein and conventional high-carbohydrate, serotonin-stimulating diets. We have identified carbohydrates with this combination of properties and call them "Good Calories."

Good Calories are very different from the carbohydrates used in male-centered serotonin-stimulating diets. These male-centered diets use high-sugar carbohydrates, which produce excess blood sugar and stimulate the release of excess insulin. Such insulin-producing carbohydrates are particularly harmful to women, so we call them "Bad Calories."

The real question concerning Good Calories is whether they cause enough insulin to be released to stimulate serotonin synthesis or whether these carbohydrates more closely resemble high-protein diets, which do not create enough insulin release for serotonin synthesis.

Five studies have examined the effect of Good Calorie foods that produce a moderate (not excessive) amount of insulin. All five say these Good Calorie foods create conditions in which serotonin synthesis is favored.[1] That is enough for a woman, whose serotonin system is designed to create serotonin more easily than a man's. This means that Good Calorie carbohydrates retain the serotonin-stimulating power traditionally associated with Bad Calories.

At the same time, Good Calories also retain the power of high-protein diets in their ability to lower insulin and inhibit appetite. Two different studies have examined the effect of Good Calories on hunger and the total amount of food you eat.[2] Both studies concluded that Good Calorie carbohydrates are better than Bad Calorie, high-insulin carbohydrates in controlling these two elements of food addiction.

In many ways, Good Calories mimic what Prozac-like drugs do when they inhibit appetite by simultaneously stimulating serotonin and decreasing insulin.[3]

Good Calorie carbohydrates also retain the insulin-lowering power of high-protein diets while adding the ability to stimulate serotonin.

All other attempts to elevate serotonin with food rely on foods that boost insulin levels so high that they elevate fat formation and create new abnormalities in brain chemistry.

A woman can use foods that produce only small amounts of insulin because it is easier for her to elevate serotonin. A woman using these lower-insulin foods does not have to worry about increasing fat formation or producing negative changes in her brain chemistry.

We use the phrase *woman-friendly foods* because these foods do almost nothing to the serotonin of a man, although they will help him to lose weight. A man's insensitivity to dietary stimulation of serotonin requires these massive surges of insulin.

This subtlety, working only with women, is why modern medicine, with its male-centered models, has ignored these foods. No other serotonin diet uses them. They are good for all women—even those who do not want to lose weight but merely want to feed their serotonin superiority.

Think of it this way. Female serotonin is like a Ferrari sports car, more responsive and quicker than any other car. Because female serotonin is so sophisticated, it requires more care than a family sedan, but the added performance is worth the effort.

Conventional diets are designed for the male serotonin system—something more closely resembling a family sedan. Conventional diets do not provide the high-octane gasoline needed to power a Ferrari. They throw a woman's body into a state of shock, causing her engine to sputter and misfire.

When a dieting woman puts the proper foods in her "tank," she responds with a new level of performance, making the superiority of her engine obvious to all. A dieting woman will no longer feel depressed or crave foods when her mind has enough serotonin and not too much insulin. She'll also feel happier, more peaceful, and more clear-headed.

But do you know what happens if you put this high-octane fuel in a family sedan? Nothing. The male engine has no way to use that added power so there is no improvement in performance.

I have *never* been on a successful diet. I have never enjoyed the thought of trying to lose weight. I have always had frightening cravings for sweets. The cravings had gotten out of hand, becoming an obsession. I could easily indulge in a pound of M&M's for lunch. Then, hide the wrapper so no one would ever know. I would also hide the four to five candy bar wrappers from my cravings throughout the day. I started this diet to eliminate my obsessive cravings. Guess what? *It worked!* I am so excited about this. *I no longer have obsessive cravings for sweets!* Plus it was not as difficult as I expected. I still ate plenty of food, so I did not feel hungry, not even once. After six weeks, I lost ten pounds and have consistently kept it off. It taught me how to properly combine foods. I now have the capability to emotionally make choices about foods, which is quite empowering. I intend to use this plan as a guide for eating, for the rest of our lives.

—*D. W., Columbus, OH*

I gained about sixty pounds [after quitting smoking]. Nothing fit, I hated myself. I found your book and it looked good. Two days later I started your diet. As of today, I have lost twenty-one pounds and [am] feeling great. I am back on the golf course again and my life is slowly, but surely getting back to normal.

—*T. B., Clearwater, FL*

TWO NEW WAYS TO USE GOOD CALORIES

The Good Calorie Diet introduced the concept of Good Calories to describe low-insulin carbohydrates that aid in weight loss. *Naturally Slim and Powerful* is an advanced version of the program contained in *The Good Calorie Diet*. It retains the weight-loss power of *The Good Calorie Diet* while maximizing its ability to alter a woman's serotonin activity.

Good Calories are the main source of carbohydrates, and Good Calorie breakfasts are our prime serotonin-stimulating meal. You eat more than Good Calories in other meals—50 percent of your calories come from carbohydrates, 25 to 30 percent from fat, and 20 to 25 percent from animal or vegetable protein.

New to our program is the concept that timing is vital. When you eat a food is as important as what foods you eat.

Another vital concept is food combining. Protein meals must be separated from the serotonin-stimulating meals. Do not eat animal protein for breakfast. The ideal breakfast would have no protein. Similarly, do not mix large amounts of high–blood sugar foods with your protein meals; those meals are for insuring a continuation of low insulin.

One of our current programs also includes caloric restriction, something forbidden in the old program. We reversed our opinion of caloric restriction because we can now take a standard caloric-restriction program and modify it so it no longer attacks a woman's serotonin.

Why do we have two programs? Wouldn't it be simpler and easier for everyone to eat Good Calories without caloric restriction? Obviously, it would, but different women require different diets.

If you are a woman who does not want to lose weight but merely wants to feed her serotonin, then you would choose the "free-eating" program.

For weight control, both our "free-eating" and caloric-restriction programs elevate your serotonin without creating insulin excesses. Which program you use depends on your body, your serotonin, and your insulin. We'll tell you how to choose.

If something other than low serotonin or high insulin created your problem you must use caloric restriction and exercise to lose weight.

However, even with caloric restriction, Good Calories are still magical. Conventional caloric restriction weight-loss programs attack a woman's serotonin. This is not a problem with our program. Furthermore, including Good Calories increases the rate of weight loss.

Therefore, the question is not whether Good Calories must be used. They must. The only question is whether unlimited amounts of Good Calories will create weight loss.

What percentage of women will lose weight without caloric restriction? To know the answer to this question, let's look at Prozac, Redux, fenfluramine, and the other diet drugs that mimic Good Calories by boosting serotonin and lowering insulin. By looking at these drugs, you can get a good idea of what Good Calories will do without caloric restriction.

Women taking these drugs spontaneously eat from 13 percent to 40 percent fewer calories without feeling hungry.[4] Not only do they want to eat less food but they no longer suffer from cravings for carbohydrates and fats.[5] This makes it easier for them to adopt a new style of eating.[6]

Just like Good Calories, these drugs produce weight loss in only some women. One third to one half of women taking these drugs will lose weight without caloric restriction.

These drugs work for the same women who will lose weight just by eating woman-friendly carbohydrates without caloric restriction.

Who are these lucky women? These are primarily the women who suffer from the symptoms of excess insulin and low serotonin—carbohydrate cravings, food bingeing, and depression when not eating.[7] They are also women with fifteen or more pounds to lose.[8] Also, women whose fat is concentrated in their abdomen tend to lose weight without caloric restriction.[9]

Women who need to lose the last few pounds, who do not suffer from excess insulin, or who suffer from fat accumulation primarily on their legs may need to combine Good Calories with caloric restriction and exercise.

The easiest way to tell if you are one of the lucky 40 percent is to spend the first ten days of your diet concentrating on Good Calories without caloric restriction. If, after the first week to ten days, you start to lose weight, then you are probably suited to the free-eating program.

If you do not find that you lose weight on the free-eating program or find that your weight loss has plateaued, or you'd like to lose the last few pounds, then you should try the program of Good Calories with caloric restriction.

Do not worry. Our program will be unlike any calorie-counting diet you have ever experienced. Good Calories will transform this old and tired method of food cravings, irritability, and depression into something new and supercharged. Woman-friendly eating makes caloric restriction easy to maintain. It is amazing how much better a woman feels when her diet no longer attacks her serotonin. The *only* difference between our two programs is whether you count calories.

> I feel as if I have had a new lease on life. I'm eating more food than I ever have, I'm eating more food that I *love* than I ever thought I could. I'm never hungry . . . I never feel the discomfort and anxiety [like on] other diets. I love this lifestyle! I am eating the way I always wanted to and never thought I could. I have lost thirteen pounds [and even] began adding wine to my dinner. Thank you, so much.
>
> —*S. D., Scarsdale, NY*

CLINICAL PROOF OUR PROGRAM WORKS

Are Good Calories and Good Calorie breakfasts powerful enough to enhance weight loss while reducing insulin? Yes. But until recently there was no clinical proof, just the experience of the tens of thousands of women who tried the original program.

However, in 1994, *The American Journal of Clinical Nutrition* published the results of the first clinical trial.[10] By accident, this trial included Good Calorie breakfasts, an essential element of *Naturally Slim and Powerful*'s program.

The study examined women who suffered from nondiabetic elevation of insulin, a prime symptom of obesity and food addiction.[11] The women were in their mid-thirties and weighed an average of slightly over 200 pounds.

For three months, the women either followed a traditional low-calorie diet or ate the same number of calories but used Good Calories for most of their carbohydrate intake.

They also separated carbohydrate-rich meals from protein-rich meals, something that lets Good Calories elevate serotonin to the maximum by allowing Good Calories to work without having to compete with proteins, which suppress serotonin synthesis.

This was not a diet that forced people to eat strange foods. A typical day for the Good Calorie group might be papaya and oatmeal for breakfast, pasta or lentils for lunch, and dinners of roast lamb with salad and vegetables. There was even a late-night snack of yogurt. Both groups got 50 percent of their calories from carbohydrates, 20 percent from protein, and 30 percent from fat.

Bad Calories were sometimes accidentally substituted for Good Calories, and other important rules of *Naturally Slim and Powerful*'s program were broken. Nevertheless, these results show clearly that Good Calories can energize the effectiveness of a traditional low-calorie diet.

Women eating Good Calories lost 22 percent more weight. On a three-month diet of approximately 1,200 calories a day, women lost an average of 20.5 pounds on Good Calories. Without Good Calories they lost an average of 16 pounds.

However, it gets even better; Good Calories produced more weight loss than the carbohydrates ordinarily used to stimulate serotonin. Good Calories cut between-meal insulin in half, preventing food addiction, and more than halved the insulin increase in the following meal.

The doctors were not sure if this meant that Good Calories produced a greater weight loss or whether they had accidentally chosen patients predisposed to lose weight. Therefore, they took the Good Calorie dieters and put them on the conventional diet. They also switched normal dieters to Good Calories. If the new Good Calorie group lost more weight in the next three months, it would prove all the magic could be attributed to Good Calories.

After three months on their new diets, *the new Good Calorie group lost 65 percent more weight* (16 pounds) and their insulin levels declined while the group that switched to a conventional diet saw their insulin levels increase and lost only 10 pounds. This proves that Good Calories do make a difference.

> When I came across your book my blood glucose level was high. I immediately cut out white bread, potatoes, [and] quick

oatmeal and tried to follow the general ideas [you] mentioned. The blood glucose [blood sugar] level came down to normal levels almost like magic.

—*G. D., Taylorville, IL*

My husband has lost eight lbs. and I've lost six. We eat more than before and are shocked to see pounds melt off. You've made me ecstatically happy.

—*P. L., North Hollywood, CA*

SIX WOMAN-FRIENDLY RULES

A woman's body is so sensitive that the environment in which she eats a food determines the way she responds to that food. The amount of insulin that a food makes changes depending on how you cook it, what foods you eat at the same time, and how the food was manufactured. This means that a woman will have a different response to the same food according to these variables. For example, if a woman is eating a diet that is extremely high in animal protein it does not matter if she eats a few woman-friendly carbohydrates; the protein inhibits her insulin production.

This means that any definition of woman-friendly foods must

SIX PATHS TO A WOMAN-FRIENDLY ENVIRONMENT

1. Never skip meals.
2. Instant foods, processed foods, overly ripe foods, overcooked foods, and saturated fats should be avoided.
3. Overall, 50 to 60 percent of calories should be from carbohydrates, 20 percent from animal or vegetable protein, and 20 to 30 percent from fat.
4. Eat only Good Calorie carbohydrates for breakfast, eat a full breakfast, and eat little or no protein for breakfast.
5. Take vitamin and mineral supplements.
6. Substitute Good Calories for Bad Calories, and cheat intelligently.

first start with a discussion of the nutritional environment that allows a food to be woman-friendly. Only after we understand the environment in which women can be slim and powerful can we go on and list the foods that are Good and Bad Calories.

We determined that six simple rules determine a woman-friendly nutritional environment. The same six rules apply to both the free-eating and caloric-restriction programs. They relax the Starvation Response and relieve the body of any fear of starvation.

As you look at these six rules, think of how poorly the normal American diet conforms to these requirements. Is it no wonder that women have so many problems with today's food?

RULE 1: *Never Skip Meals*

You must never skip a meal. The reason for this is simple: Skipping a meal implies that food is not available and makes the body think you are threatened with starvation. The body responds by elevating insulin after your next meal. Therefore, skipping meals can unleash any of the problems associated with high insulin: food addiction, excess fat formation, and abnormal serotonin.

We realize many of you skip a meal to make up for a previous dietary excess. However, this is exactly the wrong thing to do. Your previous meal, that huge meal of fat and sweets, may have set off excess insulin and food-craving mechanisms. Therefore, you must be very careful to reassure your body. The way to do this is to eat only woman-friendly carbohydrates for the next day or so. The constant flow of Good Calories will act like a medicine to reverse any damage you may have done to your levels of insulin or serotonin.

RULE 2: *Instant Foods, Processed Foods, Overly Ripe Foods, Overcooked Foods, and Saturated Fats Should Be Avoided*

One of the prime signals to set off excess insulin is eating foods that produce excess blood sugar. Unfortunately, many things can take a Good Calorie and elevate the blood sugar it produces, changing it into a Bad Calorie.

Consider rice. Overcooked rice produces up to 50 percent more blood sugar than normal rice. It is the same with any carbohydrate—the more overcooked it is, the more blood sugar and insulin it will create.

If a fruit is overly ripe it has excess sugar and will recall times of starvation, when rotting foods were the only source of sustenance. Consider bananas. A ripe banana with brown spots on the skin can produce nearly twice the blood sugar as does a green or moderately ripe banana.

Food processing also increases blood sugar. Instant potatoes, instant rice, corn chips, and puffed rice all produce more blood sugar than their unprocessed counterparts (Table 5-1). Almost all breakfast cereals are processed foods that generate high blood sugar.

Many people suffer from problems controlling their appetite because so many modern convenience foods are altered to create excess insulin. Almost all instant foods and processed foods are Bad Calories.

The type of fat in your diet also changes blood sugar. Remember that eating large amounts of animal flesh was a signal of starvation. Animal flesh contained fats as well as protein. Therefore, your body responds negatively to the presence of saturated animal fats as well as animal protein. In contrast, the presence of unsaturated vegetable fats is a signal of plenty.

Combining starches with saturated animal fats, such as butter,

TABLE 5-1
Food Processing Increases Blood Sugar

Food	Increase in Blood Sugar Compared to Unprocessed Food
Potatoes	100%
Instant potatoes	156%
Corn	100%
Corn chips	152%
Corn flakes	163%
Rice	100%
Instant rice	178%
Puffed rice	190%

increases blood sugar and insulin.[12] In contrast, combining starches with an unsaturated vegetable fat, such as olive or canola oil, decreases blood sugar and insulin.[13] The difference between unsaturated and saturated fats is so profound that a plate of pasta coated with butter (saturated animal fat) can produce four times more blood sugar than a plate of pasta coated with olive oil (unsaturated vegetable fat).

Remember that when you eat animal protein you are most likely also eating saturated fats. Unless you are eating the special seafoods that are high in unsaturated fats, you should avoid combining animal protein with starches. Try to avoid having potatoes or pasta with your meat.

The primary sources of saturated fats are meats and dairy products such as butter and cheese. Some vegetables have saturated fats, for example, avocados, nuts, and the palm and coconut oils used in candies and processed foods. Chicken and turkey also have more saturated fat than unsaturated fats, although they do have a better ratio of unsaturated to saturated fats than red meat.

Seafood often has a better ratio of unsaturated to saturated fats than red meat does, so seafood is a better choice for combining with starches. Unfortunately, not all seafood is high in unsaturated fat (Table 5-2). You can cut your consumption of saturated fat by up to 75 percent by switching from red meat to the proper seafood.

RULE 3: Overall, 50 to 60 Percent of Calories Should Be from Carbohydrates, 20 Percent from Animal or Vegetable Protein, and 20 to 30 Percent from Fat

We do not recommend going to dietary extremes. Our ancestors ate a balanced diet.

You can have meat, but only if you do not eat it for breakfast and eat it in moderation. Whenever possible, eat animal protein (meat, fish, dairy, or poultry) at lunch. This gives your body the longest time to purge itself of the proteins that inhibit serotonin synthesis before the next day's breakfast, when you intend to elevate your serotonin.

For the same reasons, it is good, whenever possible, to eat your largest meal at lunch.

If possible, your evening meal should be a carbohydrate-rich

TABLE 5-2
Unsaturated Fat in Seafood

High in Unsaturated Fats (Good)	Low in Unsaturated Fats (Bad)
clams	abalone
cod	catfish
crab	gefilte fish
haddock	herring
halibut	lobster
perch	mussels
pike, northern	oysters
salmon, Atlantic, coho, pink	pompano
scallops	roughy
sea bass	sablefish
shrimp	salmon, Chinook, chum
snapper	sea trout
striped bass	sturgeon
trout	swordfish
tuna	
whitefish	
whiting	

meal. This clears your system of animal proteins before the next morning's breakfast.

It is also a good idea to eat breakfast early and have a late lunch. This allows your morning serotonin boost to work for as long as possible before lunch interferes.

We realize the evening meal is traditionally the time when you eat your largest meal and when you eat meat. If you eat animal protein for dinner, do it as early as possible. This gives your body more time before breakfast to remove proteins that might interfere with the next morning's serotonin synthesis. You can also alternate lunch and dinner as the time when you eat your animal protein.

It is important not to have animal protein in successive meals as this will tend to inhibit serotonin by building up such high levels of protein that it will be difficult to clear them before the next morning.

Not everyone can eat animal protein and lose weight. Some women cannot restore their serotonin to normal unless they drastically reduce animal proteins or eliminate them. We found that in many cases women cannot lose weight unless they follow a vegetarian diet for at least a few weeks. It all depends on your individual sensitivity.

Do not worry about getting enough protein if you reduce your normal intake of animal protein. The U.S. government has finally decided that a vegetarian diet can provide complete and balanced nutrition. If you combine grains with half as much beans you will get all the protein you need.

At the same time, you should never think that protein is something to avoid. You must eat a balanced diet.

RULE 4: *Eat Only Good Calorie Carbohydrates for Breakfast, Eat a Full Breakfast, and Eat Little or No Protein for Breakfast*

The best time to eat Good Calories is when they stimulate the most serotonin. Eating Good Calories will do you little good if your stomach is full of other food, especially proteins that impair serotonin. Therefore, Good Calories are most powerful when they are eaten on an empty stomach, such as at breakfast.

Early morning is also the time when it is easiest to use food to stimulate serotonin synthesis.[14] Any foods eaten for breakfast will produce significantly more serotonin than the same foods eaten in the evening.[15]

Breakfast is the time when Good Calories are most needed. A woman's serotonin level is naturally low in the morning.[16] Starting the day with Good Calories will change the way she experiences the entire day. She will be much more in control and confident than when she spends the day fighting serotonin-driven food cravings, depression, and irritability. Think of Good Calorie breakfasts as the medicine that transforms the day.

Why is a woman's body so sensitive to Good Calories during the morning? Why not the evening? Once again it all goes back to the Starvation Response and the processes of evolution.

During times of plenty, food is either nearby or plentiful enough that it can be taken back to the camp and stored for consumption early in the morning. During times of starvation, a woman had to spend the day looking for food, often finding it only after a long day of searching far away from the camp. Therefore, nature has designed breakfast to determine if food is available or scarce. If there is no food, or the wrong food, then the body is already in a state of low serotonin and a woman will not recover enough to move around and waste whatever food energy is left. If a woman eats foods that signal abundance, her serotonin levels rise and she will be active, sexually receptive, and not excessively hungry.

To take advantage of this window of opportunity it is important to eat enough food for breakfast. Unfortunately, many dieting women do exactly the opposite. They often skip breakfast entirely, and most of those who do eat breakfast consume so little they stand little chance of elevating serotonin. Remember, breakfast is when a woman convinces her body that starvation has ended and food is plentiful.

However, a woman must eat *only* breakfast foods that stimulate serotonin. If you eat protein for breakfast, even a little, you will inhibit optimal serotonin synthesis. At the same time, protein sends the signal that starvation continues. Protein is best eaten later in the day.

Think of breakfast as the time to take your medicine. Instead of taking a serotonin-stimulating drug you are eating Good Calorie carbohydrates. You must eat your full dose of Good Calorie carbohydrates if you are to enjoy spontaneous weight loss and alleviate symptoms of low serotonin. Eat a *large* breakfast consisting of only Good Calories and only a small amount of fat. Eat *no* animal protein for breakfast.

However, many people are pressed for time in the morning. To make it easier for them to eat Good Calorie breakfasts, we have arranged for Good Calorie Products to market a line of Good Calorie breakfast drinks as well as baked foods, breads, cakes, and cookies. Hopefully, they will soon be available in your local grocery or health food store. If they are not yet available in your area, Appendix III tells you how to mail-order these items.

RULE 5: *Take Vitamin and Mineral Supplements*

When you are overweight or under stress you tend to excrete significantly more, and absorb fewer, vitamins and minerals.[17] The resulting deficiencies make it hard to have normal levels of serotonin and insulin. Therefore, you should take supplements. Because stress reduces these vital vitamins and minerals, it is important for all women, even those not trying to lose weight, to take supplements.

Chromium

Chromium acts on the body in the same way as do Good Calories.[18] As early as 1929, scientists observed that chromium-rich diets could decrease blood sugar.[19] In the mid-1960s, studies showed that chromium restored blood sugar to more normal levels.[20] It probably does this by making insulin work properly.[21]

Dr. Anderson of the U.S. Department of Agriculture has shown that 90 percent of Americans are so deficient in chromium that normal insulin regulation is impossible.[22] American men have only 1.9 parts chromium per million while African men have 5.5 parts per million. Near Eastern men have 11 parts per million and Far Eastern men have 15 parts per million. When it comes to chromium we are a malnourished people.

Americans are not starving, so why are they so deficient in chromium? Mental stress, food processing, and Bad Calories all induce chromium deficiencies.[23] Women also strip themselves of chromium when they go on diets that restrict food intake.[24]

Only a handful of modern foods contain significant quantities of chromium. These include brewer's yeast, beer, black pepper, liver, lobster, oysters, mushrooms, shrimp, and whole grains.

This means you should take chromium supplements. However, chromium must be in a special form before it will do anything in your body, and most chromium pills contain only inert chromium that is better for plating automotive trim.

The biologically active form of chromium is known as the glucose tolerance factor (GTF).[25] Several vitamins, especially nicotinic acid and B complex, must accompany chromium before it can act.

Of the many chromium supplements currently offered, chromium–nicotinic acid complexes appear to be the most active. Although blood sugar levels are unaffected by either chromium or nicotinic acid, combining these two produces a 15 percent decrease in blood sugar.[26] Chromium–nicotinic acid complexes were shown to decrease cholesterol by 20 to 30 percent in two different studies.[27] Chromium nicotinate is frequently sold under the brand name Chromate.

The National Research Council of the National Academy of Science recommends a dose of 50 micrograms of chromium per day. As a woman losing weight, you should probably take more than that. But please do not exceed the maximum recommended dosage of 200 micrograms per day.

Do not confuse chromium nicotinate with chromium picolinate, which recently received attention as a miracle pill to convert body fat into muscle. *Obesity & Health* magazine recently called chromium picolinate the "scam of the hour." They state that "much of the dubious information on chromium picolinate appears to originate with Nutrition 21, the California company which leases the patent, produces the compound and markets it to supplement companies." It has been reported that many of the researchers who produce favorable research articles are tied to companies that sell chromium picolinate.[28] Few of these articles have been subjected to the peer review process that ordinarily judges scientific research. Several reviews and research articles question whether chromium picolinate could increase athletic performance or muscle mass.[29] Furthermore, picolinate acid has been linked to potential causes of health problems, including altered cellular function,[30] altered cellular shape,[31] problems combining it with protein,[32] altering iron metabolism,[33] and increased excretion of minerals that may be essential for proper health.[34]

Vanadyl Sulfate

This is an active form of vanadium, a substitute for chromium. Several studies show it is more effective than chromium in controlling insulin. It is legal to sell but we are reluctant to recommend it simply because there are a few studies indicating chromosome damage may result from large doses of vanadium.

Vitamin C

Vitamin C has so much of an effect on blood sugar and insulin that earlier in this century some doctors thought that vitamin C deficiencies created diabetes.[35] Overweight people are deficient in vitamin C, which might be one reason their insulin is abnormally elevated.[36]

Vitamin D

Vitamin D levels are much lower in overweight people, a condition that can cause abnormal insulin levels.[37]

Vitamin E

Vitamin E supplements decrease insulin rushes.[38]

Magnesium

Magnesium deficiencies have also been linked to abnormal insulin levels.[39]

Potassium

In normal people, insulin stimulates potassium, but when you are overweight insulin no longer stimulates potassium.[40] This suggests that the overweight need more potassium.

Recommended Supplements

We recommend you take the following supplements: 200 micrograms of chromium nicotinate, 300 mg of vitamin C, a good B complex pill (15 mg vitamin B_1, 10 mg vitamin B_2, and 5 mg vitamin B_6), 800 IU vitamin D, and a good mineral tablet with at least the minimum RDA of calcium, zinc, potassium, and magnesium.

You should be able to fulfill your need for supplements with inexpensive products available from stores that sell vitamins. If you wish to order supplements, please see Appendix III.

RULE 6: Substitute Good Calories for Bad Calories, and Cheat Intelligently

At this point you may be wondering how you can survive on a diet of only Good Calories. Don't worry—you don't have to try. No one could survive eating only one type of food and no one should try to go through her life eating only Good Calories.

Good Calories are an enjoyable form of therapy. You eat them to feel better, but they are *not* all you will eat. The *only* time you must eat *only* Good Calories is at breakfast.

Learn how to combine Good Calories with Bad Calories. Good Calories hide a Bad Calorie by making it only a small part of a food mixture that no longer creates excess insulin and impairs serotonin. The effects of Bad Calories vanish if you eat approximately three to four times more Good Calories than Bad Calories. However, you should eat both foods in the same meal, or preferably at the same time. For example, if you have a bowl of lentil soup (consisting of Good Calories), you can use croutons composed of Bad Calorie bread crumbs.

Don't measure your food. Your intuitive feeling is probably good enough. You are probably eating enough Good Calories if you think you are eating four times more Good Calories than Bad.

There are always times where you cannot adhere to the rules of the program, even by hiding Bad Calories with Good Calories. Do not feel guilty. Cheating is an integral part of the program. Learn how to cheat well!

As we mentioned before, and it is important enough to mention again, do not try to atone for splurging by skipping your next meal. The next meal is your opportunity to restore your serotonin to normal. Take advantage of it.

Also, eat predominantly Good Calories for the next day or so. This adjusts your amino acids so you can boost serotonin at breakfast.

RULES FOR CHEATING

Before cheating:

♦ If possible, prior meals should be predominantly Good Calories

After cheating:

♦ Do not skip the next meal
♦ Eat Good Calories for the next few meals
♦ Do not feel guilty

Finally, do not feel guilty. If you have been using our program for a couple of weeks, it will take more than one or two bad meals to drive your serotonin so low that you experience any problems.

Sometimes you know ahead of time that you will be cheating. Perhaps you have been invited to a party or a late dinner. You can minimize the effects of cheating if you eat primarily Good Calories for your next few meals.

These six simple rules boost your ability to make serotonin by increasing tryptophan's ability to enter the brain, and this boost is most significant when compared with the foods dieters ordinarily eat.

High-protein diets are the current rage, yet they decrease the ability of tryptophan to enter the brain by about 30 percent.[41]

Where standard diets make their mistake is in not realizing the necessity for a carbohydrate breakfast without protein. When comparing a high-protein diet and a high-carbohydrate diet, the high-carbohydrate diet has nearly twice the serotonin-boosting power.[42]

It is easier for a woman to stimulate serotonin, especially at breakfast, and that is why women can stimulate tryptophan entry under conditions where a man would experience no increase in tryptophan entry.[43] Still, the less protein in your breakfast food, the more you will stimulate tryptophan entry. Unfortunately, the only totally protein-free carbohydrates are similar to those sold in the drinks offered by Good Calorie Products (see Appendix III). The next best thing is to choose carbohydrates with a low protein content. Table 5-3 gives the protein content of foods eaten for breakfast. (Good Calories are in bold.) Some of these carbohydrates are not commonly eaten for breakfast in our culture. Eating as low as possible on the protein scale is advisable.

Virtually every standard weight-loss diet contains enough protein to inhibit serotonin increases. We have verified that one of the most popular diet drinks, a product not generally regarded as "high protein," still contains enough protein to inhibit tryptophan entry. By simply avoiding the high levels of breakfast protein in standard diet drinks you are coming out ahead of the game. Compared to these diets, there is no question that our program boosts your serotonin-making potential.

TABLE 5-3
Protein Content of Breakfast Carbohydrates

Less Than 5% Protein	Less Than 10% Protein	More Than 10% Protein
FRUIT	FRUIT	FRUIT
apple	blackberry	apricot (11%)
apple juice	cantaloupe	casaba melon (11%)
banana (firm)	carambola	rhubarb (14%)
blueberry	cherry	
boysenberry	currant	CEREAL & VEGETABLE
crabapple	grapefruit	amaranth
cranberry	honeydew melon	beans (20%+)
date	lemon juice	breads
elderberry	lime	breads (kernel) (20%+)
fig	nectarine	buckwheat
fruit salad	orange	bulgur (12%)
grape	papaya	corn
guava	passion fruit	farina
kiwi fruit	peach	lentils
mango	raspberry	millet
pear	strawberry	oats (15%)
pineapple	tangerine	pasta (12%)
plantain (firm)	watermelon	pumpkin (13%)
plum		quinoa
pomegranate	CEREAL & VEGETABLE	rye flour
prune	barley	semolina flour
raisin	corn grits	taro
sweet potato	corn meal	wheat flour
yam	corn taco	
	hominy	READY-TO-EAT CEREAL
CEREAL & VEGETABLE	potato (firmly cooked)	All Bran (20%)
tapioca	rice	Bran Buds with psyllium (15%)
	rice bran	bran flakes
	squash (acorn)	Cheerios
	squash (butternut)	Fiber One (12%)
		Heartwise (13%)
	READY-TO-EAT CEREAL	oat bran (13%)
	Bran Chex	oatmeal (15%)
	corn flakes	Special K (19%)
	granola	wheat flakes
	puffed rice	
	Rice Chex	OTHER
		eggs
		dairy proucts
		meat

Will our program boost your serotonin in the same way as do serotonin-stimulating drugs? No, but it may let these drugs work better.

Serotonin drugs generally work by forcing your existing levels of serotonin to work harder. However, these drugs stop working unless there is enough serotonin for them to activate. These drugs are powerless once they exhaust your preexisting levels of serotonin.

This means these drugs cannot work optimally when you have a decreased ability to make serotonin. Both Prozac and Redux have been shown to stop working as you increase your levels of dietary protein, losing their ability to inhibit your appetite.[44] It does not take much to create a problem. Even eating only slightly more than an ounce of the wrong protein can inhibit fenfluramine's activity.[45] In rats, it has been shown that increasing protein content can stop Redux from inhibiting appetite.[46]

It is also the same story when you consider the serotonin antidepressants. We have already cited numerous studies that show that serotonin antidepressants stop working when you go on a diet, preventing tryptophan from entering the brain. Slightly more than an ounce of the wrong protein is all it takes to increase depression.[47]

If you are taking these drugs you should be aware that a change in diet could alter the dosage you need. Tell your doctor that you are altering your diet in a way that might increase the effectiveness of your medication.

No matter whether you take serotonin drugs or just want to eat properly, following our program will make it easier to experience your full potential.

Look at what these six rules did for these women who wrote to us, or J. C., who related her story to us over the course of several telephone consultations.

I am now using your program faithfully and feel I have some control over my life and destiny. I had a heart attack two years ago and had put on a considerable amount of weight. I had tried both low-calorie and low-fat diets with no success. My doctor kept telling me that I had to lose weight. In desperation I asked him what I should do. In all seriousness he suggested I try a "concentration camp diet"! You can only imagine the sense of desperation I felt. Needless to say, I turned to your program when all else failed, and I am so grateful I did. I have lost most of the weight I needed to and with it the taste for fats and sugar. Whenever I eat too many Bad Calories for any length of time, I notice the cravings return. But now I have an awareness of my problem and the steps to overcome it. I just return to eating lots of Good Calories and everything returns to normal. This program has given me an enormous lift, spiritually and physically, and I look and feel slim.

—*J. C., San Francisco*

I have to date lost 23.5 lbs.! It took me three months to do it and I've kept the weight off since then by doing nothing but watching the combinations of foods I eat. I pretty much eat what I want and I don't feel deprived. Your expertise and detailed information was so easy to understand and implement. I hope to lose another seven or eight lbs. but I seem to be on a plateau. I'm not worried though because I look and feel so good. I feel so good on this diet!

—*B. G., Boise, ID*

The loss of the insane craving for sourdough, sweets and sausages as well as the newly acquired appreciation of simple, unadorned food makes it all worthwhile.

—*N. B., Sausalito, CA*

CHAPTER 6

Good and Bad Calories

The key to our programs is the use of low–blood sugar carbohydrates, yet there is great controversy about this subject. Critics falsely claim that focusing on low blood sugar excludes "healthy" foods such as potatoes. These critics base their criticisms on male-centered models of human nutrition. They consider a food to be healthy if it sustains a man's body. If a food creates problems only for a dieting woman, is it healthy? It is time to extend our definition of *healthy* to include the effects on each gender and the effects on mental processes.

Furthermore, they totally ignore the reality of our program. We do not *exclude* any carbohydrate. Although it is true that a baked potato produces excessive amounts of blood sugar and insulin, it is not excluded from our programs.

Our use of *only* low–blood sugar carbohydrates is confined *only* to breakfast. That is the time when a woman's body is most sensitive, so she must be careful to feed it low-sugar foods that resemble ancient signals of abundance.

Lunch and dinner are not as stringent. One meal is animal protein oriented and low in carbohydrates. This allows a woman to eat animal protein without excessively elevating her insulin. The other meal is a combination of high– and low–blood sugar carbohydrates and other fruits and vegetables necessary to balance the diet. Of course, you can follow a vegetarian alternative.

In no way does our program exclude nutritious foods. It just says certain other factors are important, such as how a food is cooked, when it is eaten, what is eaten at the same time, and how the food is processed.

There was another erroneous criticism of low–blood sugar carbohydrates. It was said that mixing low– and high–blood sugar foods will block the benefits of the low–blood sugar foods, making them vanish in a pool of blood sugar and insulin created by the high–blood sugar foods. However, several doctors have recently demonstrated that this criticism was based on faulty experimental methodology. Good Calorie foods will lower blood sugar even if you eat other foods at the same time.

Furthermore, because breakfast is a meal of *only* low–blood sugar carbohydrates, it does not matter if high–blood sugar foods or protein could block the effects of low–blood sugar foods. In this critical meal there are no other foods to worry about.

AN INTUITIVE FEEL FOR GOOD AND BAD CALORIES

Now it's time to see which foods are low–blood sugar Good Calories and high–blood sugar Bad Calories.

There are two ways to identify Good Calories. The best way is to develop an intuitive feel for what increases blood sugar and insulin. You can also look at a chart detailing experimental results. That is tedious.

Therefore, we will start by looking at some common carbohydrates and see what it is that makes them into Good or Bad Calories. Hopefully this will help you develop an understanding of what foods are Good Calories and why they are good.

In general, a food that is high in gummy fiber is a Good Calorie. Foods that are ground in fine particles or high in sugar are generally Bad Calories.

Breads, Cookies, Cakes, and Pastries

Modern wheat flour is ground into fine particles that resemble sugar in their effect on the body. That is what makes most breads, cakes, cookies, and pastries melt in your mouth like sugar and creates excess blood insulin.

The easiest way to tell if a bread is a Bad Calorie is to place a piece in your mouth and see if it dissolves without being chewed. Only wheat that has been finely ground dissolves in that fashion. Unfortunately, almost all bakery items fail this test. Low-fat "diet" breads, muffins, and cookies produce even greater sugar and insulin rushes.

There are two ways to make Good Calorie baked goods: by using whole, unprocessed *kernels* of grain (wheat, rye, or barley "berries"), grinding the grain into large-sized particles,[1] or adding a viscous fiber, such as guar gum.[2] Appendix IV gives recipes of several breads, cakes, and cookies that use guar gum to lower their blood sugar.

For those of you who are not bakers, we are trying to arrange for someone to market a line of Good Calorie baked breads, cakes, and cookies (see Appendix III for how to obtain information on any possible products). These items are for your convenience only. You can duplicate them by following our recipes or simply substituting other Good Calorie carbohydrates.

Unfortunately, almost *all* commercially available breads are Bad Calories. The only commercially available Good Calorie breads are hard German rye breads and breads with over 75 percent of their content comprising *whole* kernels of grain (sometimes called wheat or rye berries). You can recognize these breads because they are sliced thin and very chewy.

Pasta

The good news is that pasta is a Good Calorie. It appears that the cold excursion process used to make pasta also converts wheat flour into a Good Calorie. To insure that pasta retains its special properties, make sure not to overcook it.

Canning doubles the blood sugar potential of pasta. Even if you do not eat the many canned pastas designed for children, be careful of canned soups that contain pasta, such as chicken-and-noodle soup.

Egg-enriched pasta creates low amounts of blood sugar, but it should be avoided for breakfast because it contains animal protein.

Potatoes, Yams, and Sweet Potatoes

White potatoes do not resemble the roots our ancestors ate. They have been bred specifically for sugar content. This makes them into Bad Calories.

Yams and sweet potatoes more closely resemble ancient root vegetables. They are preferable to white potatoes for this reason.

When we lectured in Ireland people constantly asked about how to eat white potatoes, the staple of the Irish diet. We replied that if you are going to eat a food that is high in sugar, such as potatoes, you must make sure not to overcook it. Potatoes that are still slightly crisp and firm will release less blood sugar than potatoes that are overcooked and melt under your fork.

Another trick is to combine potatoes with olive oil rather than butter. That will decrease the blood sugar produced; however, you have to be careful not to eat too much oil.

Instant or canned potatoes release too much blood sugar. Avoid them. Many fast-food restaurants manufacture their french fries from instant potatoes. Also, do not eat french fries that have been cooked in saturated animal fats.

Rice

Most rices are Good Calories, but instant rice is a Bad Calorie. Good Calorie rice requires about twenty minutes to cook. The process of making rice into an instant food destroys the fibrous structure needed for Good Calories.

Some rices, such as basmati and jasmine, create extra blood sugar; of course, that is why they taste so good. If you want a general rule for determining if a rice is low blood sugar, go with the short-grained varieties.

Other Bad Calorie rices include sweet dessert rices, such as the glutinous rice found in Thai restaurants, the overcooked rice found in some Chinese and Japanese restaurants, and the overcooked rice found in many convenience foods, frozen and packaged. However, most rice found in restaurants is probably a Good Calorie.

Rice Cakes and Rice Drinks

Rice cakes are extremely processed. In no way do puffed kernels of rice cakes resemble properly cooked rice. They release about twice the blood sugar as properly cooked rice.

Rice drinks are becoming increasingly popular because they provide one of the strongest known sugar rushes. Avoid them.

Corn

The corn used by Native Americans bears no resemblance to modern corns. Ancient corn was only a few inches long and contained almost no sugar. Modern corn creates high amounts of blood sugar. Avoid processed corn products, such as corn chips and corn flakes. Polenta and corn tortillas should be used sparingly and not for breakfast.

Breakfast Cereals

Most are so processed they are Bad Calories. Even the low-fat cereals should be avoided.

A few national brands are Good Calories, such as Kellogg's All Bran and Fiberwise cereals. Kellogg's Special K and Sultana Bran are marginally acceptable. General Mills' Fiber One is also a Good Calorie. Use milk sparingly, as only a small quantity of animal protein is acceptable for breakfast on occasion.

Slow-cooked oatmeal appears to be one of the best choices for breakfast and forms the foundation of many Good Calorie breakfasts. Instant oatmeals are to be avoided, and quick-cooking oatmeal (five to six minutes on the stovetop) is marginally acceptable.

Fruit

Our ancestors ate fruits immediately after picking them from the plant. They were not overly ripe. Most of the conversion of carbohydrates into the sugar that gives modern foods their ripeness takes place after they are picked. Overly ripe fruit should be avoided because it creates excess blood sugar. For example, avoid overly ripe bananas, but firm bananas are fine.

Many modern melons are new foods bred for sugar content. It is best to avoid them. Among the worst offenders are watermelon and honeydew melon.

However, many fruits are very close to the foods of our ancestors. Among the best are cherries, peaches, grapes, grapefruit, and apples.

Fruits are great snacks and breakfast items.

Vegetables

Most vegetables are Good Calories, which is not surprising because they closely resemble our ancestors' preferred diet. The most notable exceptions are carrots and potatoes, which have a higher sugar content. It is best to use carrots or potatoes in smaller portions combined with other Good Calorie vegetables to lower the total blood sugar produced by the mixture. Do not overcook them.

Juices

Juicing a fruit or vegetable creates more blood sugar because it destroys plant fibers. Some juices, such as carrot and beet, give more of a sugar rush than soda pop. A few juices, such as apple, can be acceptable if they are not made from high-sugar apples. Orange juice is also a Good Calorie if it does not have added sugar or fructose.

Legumes (Beans)

Legumes produce very little blood sugar. Combining legumes with other foods drastically reduces the amount of blood sugar produced by the resulting meal.

Eating a portion of legumes with lunch and dinner lowers the blood sugar produced by each meal. It also allows you to cheat on the ingredients in the rest of your meal. One trick is to start meals with bean soup or serve beans with entrees. One of our favorite foods is an Australian yellow bean pasta, which is found in health food stores.

Combining one part grain, such as rice, with half a part legume (bean) produces a mixture of amino acids that is a perfect substitute for animal protein.

Fructose

The most commonly used sweetener in manufactured foods is corn syrup. Fructose, the active ingredient in corn syrup, forms more of the American diet than table sugar. The average American consumes 15 percent of his or her calories as fructose.[3]

You would think that fructose, because it has a low glycemic index, is a Good Calorie. However, this is far from the truth. Of the many forms of sugar in our diet, fructose is the easiest for the body to convert to fat.[4] Combined with fat, such as in a candy bar, fructose is capable of producing more new fat than common table sugar.[5]

Fructose increases insulin levels and insulin resistance in people predisposed to high levels of insulin or blood sugar.[6] This means that fructose will force a woman toward the high insulin associated with food addiction. We recommend that all women avoid fructose, high-fructose corn syrup, and corn solids.

GOOD CALORIE BREAKFASTS

What is a Good Calorie breakfast?

Good Calorie foods for breakfast might include long-cooking oatmeal, toasted bread (hard rye, barley kernel, wheat berry), rice, yams, fruit, pasta, rice, or the few cereals that create low blood sugar.

Unfortunately, these are not the foods that people in America and Europe are accustomed to eating for breakfast. Most women either eat a traditional breakfast of high-protein eggs or Bad Calorie carbohydrates such as cereal, breads, muffins, bagels, or breakfast bars. Table 6-1 lists the most common Good and Bad Calorie foods that people eat for breakfast.

We have included breakfast cereals as recommended breakfasts. However, the inclusion of milk lowers the ability of cereals to make serotonin. Therefore, you should eat them only occasionally. We realize this involves a great change in your eating habits, but also remember your current eating habits contribute to serotonin imbalances.

Remember, you must eat enough food at breakfast to create sufficient serotonin. Just eating a breakfast of 100 or 200 calories is not enough.

TABLE 6-1
Breakfast Foods

Good Breakfasts	Bad Breakfasts
oatmeal, long cooking	eggs
Sultana Bran cereal	breakfast meats
Kellogg's All-Bran	bagels
cracked wheat, wheat berry, barley, or hard rye bread	yogurt
rice, sweetened if desired	raisins
grainy pumpernickel bread	bread, most types
Kellogg's Fiber One cereal	breakfast cereals (flakes, puffed, shredded, or granola)
applesauce, unsweetened	overly ripe fruit
orange juice, unsweetened	juice, sweetened
pasta	nuts
yams	most muffins
most fruits, not overly ripe	honeydew melon

In order to make it easier to eat the proper breakfast foods, we have arranged with Good Calorie Products to offer a line of breakfast foods. These foods include breakfast drinks, and may expand to include breakfast bars, pancakes, and cookies (see Appendix III). You do not have to eat these foods in order for the program to work. But they are easier, more convenient, and make the program easier to implement.

THE GLYCEMIC INDEX

An intuitive feel for Good Calories will help you as you go about your daily life. Few people are going to memorize tables of Good and Bad Calories. However, it is important to know that scientists have devised a number they call the "glycemic index," which quantifies how much blood sugar a carbohydrate makes.

A higher number on the glycemic index means a food produces more blood sugar. A food with a glycemic index of 90 will produce three times the blood sugar as a food with a glycemic index of 30.

WHAT FOODS ARE GOOD CALORIES?

Where is the dividing line between Good and Bad Calorie carbohydrates?

Philip's last book used the phrase *Good Calories* to describe carbohydrates with a glycemic index lower than table sugar, below 80. While this was an arbitrary dividing point, it seemed to work for the tens of thousands of women who lost weight.

Table 6-2 gives a partial list of Good Calorie foods; hundreds more are listed in Appendix I, and a list of both Good and Bad Calorie foods along with their glycemic indexes is given in Appendix II. It is not necessary that you pay attention to the glycemic index of each food. Just choose Good Calories for breakfast. In other meals make sure that you have more Good Calories than Bad Calories.

Just because a food is not given in this list does not mean that it is known to be a Bad Calorie. This table includes only those items whose blood sugar–forming ability has been measured experimentally.

The values of some foods, such as sugar, vary slightly from those we gave in Philip's last book. These changes reflect the newest data.

TABLE 6-2
Good Calorie Carbohydrates

Bakery Products
Breads
Barley kernel bread

80% kernels	54
50% kernels	66
Oat bran bread	
50% oat bran	63
45% oat bran	72
Rye kernel bread	
80% kernels	66
Hard pumpernickel	58
Whole-grain pumpernickel	66
Bulgur bread (cracked wheat)	
75% cracked wheat kernels	69

Breakfast Cereals

All-Bran (Kellogg's)	43
Bran Buds with psyllium (Kellogg's)	67
Oat bran, raw	78
Oats, long cooking	49
Rice bran, raw	27
Special K (Kellogg's)	77
Sultana Bran (Kellogg's)	74

Cereal Grains

Barley	
Barley	35
Barley, pearled	37
Cracked	72
Buckwheat	78
Bulgur, boiled 20 minutes	68
Specialty rices	
Cajun Style (Uncle Ben's)	72
Garden Style (Uncle Ben's)	79
Long Grain & Wild (Uncle Ben's)	77
Rice, brown	79
Rice, white	
long grain	73
parboiled rice	68
Uncle Ben's converted, boiled 20–30 minutes	54
Uncle Ben's converted, long grain, boiled 20–30 minutes	64
Rye, whole kernel	48
Wheat, whole kernel	59
Wheat kernels, quick cooking	75

Crackers

Fruit and Fruit Products

Apple	53
Golden Delicious	56
Braeburn	45
Apple juice, unsweetened	57
Applesauce, unsweetened	41
Apricots, dried	44

Banana, underripe	43
Cherries	32
Fruit cocktail, canned in fruit juice (Delmonte)	79
Grapefruit	36
Grapefruit juice, unsweetened	69
Grapes	62
Kiwifruit	75
Orange	62
Orange juice	74
Peach, fresh	40
Peach, canned in natural juice	43
Peach, canned in light syrup	74
Pear	51
Plum	34

Legumes

Black-eyed peas, dried	59
Butter beans	44
Chickpeas, canned	60
Chickpeas, dried	47
Kidney beans, canned	74
Kidney beans, dried	49
Lentils, green, canned	74
Lentils, green, dried	42
Lentils, red, dried	36
Lima beans, baby, frozen	46
Lima beans, dried	36
Navy beans, white, dried	54
Pinto beans, canned	64
Pinto beans, dried	55
Romano beans	65
Soy beans, canned	22
Soy beans, dried	20
Split peas, yellow, dried	45

Pasta

Capellini	64
Fettuccine, egg-enriched	46
Linguine	68
Macaroni	64
Spaghetti, durum	78

Pasta *(continued)*

Spaghetti, protein enriched	38
Spaghetti, white	56
Spaghetti, whole meal	61
Star pasta	54
Vermicelli	50

Root Vegetables

Potato, new, undercooked	67
Sweet potato	77
Yam	73

Most other root vegetables are Bad Calories

Vegetables

Peas, dried	56
Peas, green	77
Peas, green, frozen	68

Most other vegetables are Good Calories

Sugars

Fructose (see warning against use, p. 85)	32
Lactose	65

Soups

Lentil, canned	63
Tomato	54

Ethnic Foods and Meals

Pima Indian Foods

Acorns, stewed with venison	23
Corn hominy	57
Lima-bean broth	51
Mesquite cakes	36
Tortilla from crudely ground grain	54
White teparies broth	44
Yellow teparies broth	41

South African Foods

Brown beans	34
Gram dal	7

Mexican Foods

Black beans	43
Brown beans	54
Nopal, prickly pear cactus	10

Asian Indian Foods

Baisen chapati	39
Bajra chapati	82
Barley chapati	61
Bengal gram dal, chickpeas	16
Black gram	61
Black gram dal	66
Green gram	54
Horse gram	73
Rajmah	27
Semolina, steamed	79

Australian Aboriginal Foods

Blackbean seed	11
Bunya nut pine	67
Bush honey, sugar bag	61
Cheeky yam	49
Mulga seed	11

Pacific Islands Foods

Sweet potato	63

Chinese Foods

Lungkow bean thread	37

WHAT FOODS ARE BAD CALORIES?

Almost everything that women currently eat for breakfast is either protein or a Bad Calorie. Look at Table 6-3. Most breakfast cereals, breads, and muffins produce excess insulin.

Many other low-calorie, low-fat "diet" foods are insulin disasters: watermelon, instant rice, instant oatmeal, sweet corn, rice cakes, carrot juice, ripe bananas, and baked potatoes, to name a few.

Faced with a diet composed largely of Bad Calories, most women are constantly attacking their serotonin. No wonder today's women suffer so much.

Even worse are conventional serotonin-boosting diets, which seem to choose their ingredients almost exclusively from this list of Bad Calories.

Do not become worried when you look at this list. You do not have to totally give up Bad Calories; just avoid them for breakfast. The rest of the time you can use Good Calorie foods to suppress the excess insulin of Bad Calories. When properly combined with Good Calories, you can eat Bad Calories for meals other than breakfast.

Just because a food is not given in this list does not mean it is not a Bad Calorie. This table includes only those items whose blood sugar–forming ability has been measured experimentally.

TABLE 6-3
Bad Calorie Carbohydrates

Bakery Products

Cake	
Angel food	95
Flan	93
Croissant	96
Crumpet	98
Doughnut	108
Muffins	
Corn, low-amylose (sweet corn)	146
Oatmeal muffin mix (Quaker Oats)	98
Pizza, cheese (Pillsbury)	86
Waffles (Aunt Jemima, Quaker Oats)	109

Beverages

Cordial, orange	94
Lucozade	136
Soft drink, Fanta (Coca-Cola Bottlers)	97

Breads

Bagel, white	103
Barley flour bread	95
Bread stuffing, Paxo (Campbell Soup Co.)	106
Bulgur bread (cracked wheat)	
50% cracked wheat kernels	83
French baguette	136

Hamburger bun	87
Kaiser roll	104
Melba toast	100
Rye flour bread	92
Whole-meal	89
Rye kernel bread	
Cocktail, sliced	84
Wheat bread	
White flour	101
White flour, high-fiber	97
Wheat bread, gluten-free	
Pita bread, white	82
Semolina bread	92
Whole-meal flour	100
Whole wheat snack bread (Ryvita)	105

Breakfast Cereals

Bran Buds (Kellogg's)	83
Bran Chex (Kellogg's)	83
Cheerios (General Mills)	106
Cocopops (Kellogg's)	110
Corn Bran	107
Corn Chex	118
Corn Flakes (Kellogg's)	110
Corn Flakes	122
Cream of Wheat (Nabisco)	94
Cream of Wheat, instant (Nabisco)	105
Crispix (Kellogg's)	124
Golden Grahams (General Mills)	102
Grape-Nuts (Post)	96
Grape-Nuts Flakes (Post)	114
Life (Quaker Oats)	94
Müeslix	94
Nutri-grain (Kellogg's)	94
Oatmeal, quick	
One Minute Oats (Quaker)	94
Quaker Quick Oats (Quaker)	93
Oats, porridge	89
Oats, rolled	85
Pro Stars (General Mills)	102
Puffed rice	132

Breakfast Cereals *(continued)*

Puffed Wheat (Quaker Oats)	96
Rice Bubbles (Kellogg's)	126
Rice Chex (Nabisco)	127
Rice Krispies (Kellogg's)	117
Shredded Wheat	
Mini Wheats (Kellogg's)	83
Shredded Wheat (Nabisco)	118
Sustain (Kellogg's)	97
Team (Nabisco)	117
Total (General Mills)	109
Wheat, flaked	
Vita-Bits	87
Wheetabix	106

Cereal Grains

Barley, rolled	94
Cornmeal	98
Taco shells (Old El Paso)	97
Couscous	93
Millet	101
Rice, white	
Basmati	83
Instant	128
High amylose	91
Low amylose	83
Waxy	126
Specialty rices	
Mexican Fast and Fancy (Uncle Ben's)	83
Tapioca, boiled with milk	115

Cookies

Digestive biscuits	82
Graham Wafers (Nabisco)	106
Shortbread	91
Vanilla wafers	110

Crackers

Rice cakes	117
Rye crispbread, high-fiber	
Kavli Norwegian Crispbread	101
Ryvita (Ryvita Co.)	84
Rye crispbread	90
Stoned Wheat Thins (Nabisco)	96
Water crackers	90
Premium Soda Crackers (Nabisco)	106

Fruit and Fruit Products

Apricots, canned, light syrup	91
Banana	91
Mango	80
Papaya	83
Peaches, canned, heavy syrup	83
Pineapple	94
Raisin	91
Rock melon	93
Sultana	80
Watermelon	103

Legumes

Baked beans, canned (Libby's)	80

Pasta

Macaroni & Cheese, boxed (Kraft)	92
Rice pasta, brown, low-amylose rice flour	131

Root Vegetables

Beets	91
Carrots	101
Parsnip	139
Potato, baked without fat	121
Potato, french fried	107
Potato, instant	118
Potato, microwaved	117
Potato, new, well cooked	100
Rutabaga	103

Vegetables

Pumpkin	107
Sweet corn	82

All other vegetables are Good Calories.

Sugars

Honey	104
Glucose	138
Maltose	158
Sucrose (table sugar)	84

Snack Foods and Confectionery

Jelly beans	114
Life Savers (Nestle)	100
Mars Bar	97
Muesli Bar	87
Corn chips	105

Soups

Black Bean (Wil Pak)	92
Green Pea, canned (Campbell's)	94
Split Pea (Wil Pak)	86

Ethnic Foods and Meals

Asian Indian Foods

Green gram dal	89
Green gram, whole	81
Jowar	110
Lentils & Rice (East Indian Foods, frozen)	81
Maize, chapati	89
Ragi	123
Semolina, preroasted	109
Tapioca, steamed	100

Chinese Foods

Chinese glutinous rice	98
Rice vermicelli	83

Pima Indian Foods

Cactus jam	130
Fruit leather	100

South African Foods

Maize meal porridge, unrefined	101
Maize meal porridge, refined	106
M'fino, wild greens	97

I am forty-two years old and have had a weight problem most of my adult life. I lost twenty-three pounds [in three months on your program] and even during Christmas and "splurging" a little I didn't gain any weight back! Your program is a healthier way to live and look at food.

—*B. A., James Store, VA*

To my amazement, your food program works. I no longer have the craving for food that I had before trying your program. I am seventy-one years of age. I am diabetic, insulin dependent. The best part of this plan is the fact that, as long as I eat the prescribed foods, I do not have to take insulin. Feeling better, with more energy is an absolute God send, for diabetics are always sleepy and hungry. I am a registered nurse.

—*A. R., Hohenwald, TN*

CHAPTER 7

Free-Eating Program

All that remains is for you to choose between the free-eating program and the caloric-restriction program. Each has its advantages. Remember, the free-eating program is designed for women who need to lose more than fifteen pounds, mostly from their abdomen, and who also suffer from excess insulin and blood sugar.

If you are not sure which program to choose, try the free-eating program for ten days and then ask yourself if you are experiencing less food cravings *and* are starting to lose weight. If the answer is yes, continue with the unrestricted program until it stops working and then move to the caloric-restriction program.

HAVE REASONABLE EXPECTATIONS

Please be reasonable about what you want from a diet. We do not recommend losing more than two pounds per week. If you lose weight any faster your body will go into emergency mode even if you are eating Good Calories.

Furthermore, be reasonable about how much weight to lose. Serotonin-elevating drugs do not turn women into anorexic models whose bodies resemble teenage girls barely into puberty. In the same way, the *Naturally Slim and Powerful* program is not designed to produce unhealthy weight loss. It is designed to return women to a full feeling of well-being while allowing them to have the bodies they would have enjoyed had they not suffered through Bad Calories and conventional diets.

If you want to be extremely thin you probably will have to use caloric restriction. Please understand if you make yourself too thin you will have a hard time maintaining the proper brain chemistry. You will have to be very vigilant about eating only Good Calories if you want to feel good and think clearly.

The Metropolitan Life Insurance Company's 1983 table of ideal body weights gives a good idea of the range of weights women's bodies will enjoy when they eat properly. These tables show the weight that gives women the best health, not what makes them look like underweight teenage models. To go significantly below the lower end of these scales generally involves caloric restriction.

Metropolitan Life 1983 Table of Ideal Weights
(in Pounds)

Height	Small Frame	Medium Frame	Large Frame
4'9"	99–108	106–118	115–128
4'10"	100–110	108–120	117–131
4'11"	101–112	110–121	119–134
5'	103–115	112–126	122–137
5'1"	105–118	116–129	125–140
5'2"	108–121	118–132	128–144
5'3"	111–124	121–135	131-148
5'4"	114–127	124–138	134–152
5'5"	117–130	127–141	137–156
5'6"	120–133	130–144	140–160
5'7"	123–136	133–147	143–164
5'8"	126–139	139–150	146–167
5'9"	129–142	139–153	149–170
5'10"	132–145	142–156	152–173
5'11"	135–148	145–159	155–176

FREE-EATING MEAL PLAN

This section gives you four weeks of meal plans for the free-eating diet. All breakfasts are Good Calorie meals. One of the two remaining meals features animal protein. If you want to be vegetarian you can just apply meals that feature vegetables, grains, legumes, or pasta. By properly using Good Calories you can enjoy a wide variety of familiar and satisfying meals.

The next chapter gives the same meals, but with portions for a 1,200-calorie-a-day diet. You should use this meal plan only if necessary.

These meal plans are intended to be helpful guides for you to see what kinds of foods and what combinations of foods to eat. They are not an absolute regimen that must be adhered to. It is a guide to help you plan your own serotonin-enhancing meals.

Remember, it is permissible to switch lunch and dinner menus so that you have animal protein for dinner occasionally.

Also remember that some women find a vegetarian diet works best; these women should just use the vegetable meal plans.

Remember that any eating plan should follow the six rules of a woman-friendly nutritional environment.

SIX PATHS TO A WOMAN-FRIENDLY ENVIRONMENT

1. Never skip meals.
2. Instant foods, processed foods, overly ripe foods, overcooked foods, and saturated fats should be avoided.
3. Overall, 50 to 60 percent of calories should be from carbohydrates, 20 percent from animal or vegetable protein, and 20 to 30 percent from fat.
4. Eat only Good Calorie carbohydrates for breakfast, eat a full breakfast, and eat little or no protein for breakfast.
5. Take vitamin and mineral supplements.
6. Substitute Good Calories for Bad Calories, and cheat intelligently.

SAMPLE DAILY MENUS

Items followed by an asterisk are taken from our recipe section (Appendix IV).

DAY 1

Breakfast
long-cooking oatmeal
Good Calorie Bran Muffin*
fresh orange
coffee or tea (Milk may be
    added. However, keep milk
    consumption to a minimum
    for breakfast.)

Lunch
Pasta with Asparagus and
    Roasted Peppers*
green salad with Low-Fat Blue
    Cheese Dressing*
fresh apple

Snack
plain nonfat yogurt sweetened
    with 1 tsp. of table sugar or
    artificial-sweetener equivalent

Dinner
Lentil and Barley Pilaf*
steamed broccoli with Parmesan
    cheese

DAY 2

Breakfast
2 slices Good Calorie
    Applesauce Bread*
Apple Butter*
orange juice
coffee or tea

Lunch
Roasted Vegetable and Pasta
    Salad*
Spinach and Red Onion Stirfry*
green salad with olive oil and
    lemon dressing

Snack
apple juice
low-fat cottage cheese

Dinner
Yellow Split Pea Soup*
Rice Salad with Vegetables and
    Herbs*
fresh peach

DAY 3

Breakfast
baked yam with butter
fresh tangerine
coffee or tea

Lunch
Creamy Rice and Beans*
steamed spaghetti squash with
    Parmesan cheese
steamed artichoke

Snack
Good Calorie Oat Muffin*
fresh cherries

DAY 3 *(continued)*
Dinner
Asparagus Soup*
Italian Frittata*
spinach salad with Garlic and
    Buttermilk Dressing*

DAY 4

Breakfast
2 Good Calorie Apricot–Oat Bran
    Muffins*
fresh apple
coffee or tea

Lunch
lentil soup
green salad with olive oil and
    vinegar

Snack
fresh fruit salad

Dinner
Vegetable Stirfry*
rice

DAY 5

Breakfast
long-cooking oatmeal
fresh blueberries
coffee or tea

Lunch
Wheat Berry Soup*
Tabouli*
green salad with Cottage Cheese
    Dressing*

Snack
Dry-roasted Curried Chickpeas*

Dinner
Spinach Lasagna*
steamed green beans with olive oil
fresh peach

DAY 6

Breakfast
All-Bran cereal (Kellogg's)
low-fat milk
coffee or tea

Lunch
Vegetable Terrine*
brown rice
Asian pear

Snack
low-fat cottage cheese
celery sticks

Dinner
pasta topped with olive oil and
    Parmesan cheese
steamed peas
spinach salad with Low-Fat Blue
    Cheese Dressing*

DAY 7

Breakfast
2 slices Good Calorie Cracked
    Wheat Bread*
butter
coffee or tea

Lunch
plain pinto beans
rice
salsa
corn tortillas

Snack
Eggplant Dip*
raw vegetables

Dinner
Fresh Basil and Tomato Pasta*
green salad with Yogurt
    Dressing*
fresh apple

DAY 8

Breakfast
long-cooking oatmeal
fresh strawberries
orange juice
coffee or tea

Lunch
grilled chicken breast
string beans
green salad with olive oil and
    feta cheese

Snack
fresh apple

Dinner
Gazpacho*
Pad Thai*

DAY 9

Breakfast
2 slices German hard rye bread
Apple Butter*
fresh grapefruit
coffee or tea

Lunch
Steamed Flounder with Herbs*
Margrit's Curried Brown Rice*
green salad with Garlic and
    Buttermilk Dressing*

Snack
Spicy Bean Dip*
raw vegetables

Dinner
Asparagus Soup*
Roasted Vegetable and
    Pasta Salad*

DAY 10

Breakfast
baked yam with butter
fresh peach
coffee or tea

Lunch
Lentil and Spinach Stew*
turkey sandwich with lettuce,
    tomato, and mustard on
    Good Calorie Sauerkraut-
    Rye Bread*
fresh strawberries

Snack
fresh strawberries

DAY 10 (*continued*)
Dinner
Fresh Basil and Tomato Pasta*
steamed zucchini

DAY 11

Breakfast
2 Good Calorie Bran Muffins*
fresh grapefruit
coffee or tea

Lunch
Shrimp and Pea Pod Stirfry*
rice

Snack
fresh apricots

Dinner
Black Bean Chili*
green salad with Garlic and
    Buttermilk Dressing*

DAY 12

Breakfast
All-Bran cereal (Kellogg's)
low-fat milk
fresh strawberries
coffee or tea

Lunch
lamb steak
steamed broccoli with lemon
green salad with Creamy
    Thousand Island Dressing*
fresh figs

Snack
Sweet Potato Fries*

Dinner
Lentil and Barley Pilaf*
cucumber salad with vinaigrette

DAY 13

Breakfast
long-cooking oatmeal
firm banana
coffee or tea

Lunch
Black Bean and Crab Bisque*
Vegetable–Cottage Cheese Salad*
slice of Good Calorie
    Sauerkraut-Rye Bread*

Snack
Eggplant Dip*
raw vegetables

Dinner
Poached Fish with White Wine
    and Fennel*
Summer Squash Soufflé*
green salad with Yogurt
    Dressing*

DAY 14

Breakfast
2 slices Good Calorie Oat
    Bread*
fresh peach slices
coffee or tea

Lunch
pasta with tomato sauce
Sesame Green Bean Salad*

Snack
Dry-roasted Chickpeas*

Dinner
steamed vegetables
low-fat tofu
Tamari-Ginger Sauce*
rice
Cucumber, Tomato, and Yogurt
    Salad*

DAY 15

Breakfast
2 slices Good Calorie Cracked
    Wheat Bread*
unsweetened strawberry jam
orange juice
coffee or tea

Lunch
Steamed Flounder with Herbs*
new potatoes
green salad with Cottage Cheese
    Dressing*

Snack
Curried Yogurt Dip*
raw vegetables

Dinner
Quick Garlic Soup*
Middle Eastern Lentils and Rice*

DAY 16

Breakfast
long-cooking oatmeal
fresh nectarine
coffee or tea

Lunch
tomato soup
grilled coho salmon
steamed kale

Snack
hot carob drink (hot milk with
    carob powder)

Dinner
steamed vegetables
barley
Tamari-Ginger Sauce*
unsweetened applesauce

DAY 17

Breakfast
long-cooking oatmeal
fresh raspberries
coffee or tea

Lunch
grilled chicken breast
steamed zucchini and yellow
    squash
green salad with Creamy
    Thousand Island Dressing*

Snack
Sweet Potato Fries*

DAY 17 *(continued)*
Dinner
Quick Egg Drop Soup*
Spinach Fettuccine with
    Sun-dried Tomatoes*
steamed asparagus

DAY 18

Breakfast
steamed yam with butter
fresh grapes
coffee or tea

Lunch
Shrimp with Spicy Black Bean
    Sauce*
rice
fresh apple

Snack
Curried Yogurt Dip*
raw vegetables

Dinner
Quick Vegetable Soup*
Spinach Lasagna*
green salad with Garlic and
    Buttermilk Dressing*

DAY 19

Breakfast
long-cooking oatmeal
orange juice
coffee or tea

Lunch
Lemon Chicken*
steamed broccoli
cucumber salad with olive oil
    vinaigrette

Snack
1 slice of Good Calorie Oat Bread*
low-fat cottage cheese

Dinner
Chickpea Salad with Tomatoes
    and Vinaigrette*
steamed artichoke
fresh apple

DAY 20

Breakfast
2 slices Good Calorie Cracked
    Wheat Bread*
Apple Butter*
coffee or tea

Lunch
broiled halibut
steamed cauliflower
broiled tomato with Parmesan
    cheese
spinach salad with
    Yogurt Dressing*
fresh plum

Snack
Eggplant Dip*
raw vegetables

Dinner
steamed vegetables
low-fat tofu
Tamari-Ginger Sauce*
barley

DAY 21

Breakfast
All-Bran cereal (Kellogg's)
low-fat milk
fresh strawberries
coffee or tea

Lunch
1 slice Good Calorie
    Sauerkraut-Rye Bread*
Black Bean and Crab Bisque*
green salad with olive oil
    vinaigrette

Snack
fresh plum

Dinner
Lemon-Bulgur Timbales*
Sumptuous Broccoli*
baked yam with salsa

DAY 22

Breakfast
2 Good Calorie
    Apricot–Oat Bran Muffins*
unsweetened strawberry jam
coffee or tea

Lunch
Elegant Fish Pâté*
steamed asparagus
green salad with Creamy
    Thousand Island Dressing*

Snack
fresh blueberries
vanilla low-fat ice cream

Dinner
Pasta with Fresh Vegetables*

DAY 23

Breakfast
2 slices Good Calorie
    Applesauce Bread*
fresh grapes
coffee or tea

Lunch
Hummus* sandwich with
    lettuce, tomato, and sprouts
2 slices Good Calorie Cracked
    Wheat Bread*
green salad with Low-Fat Blue
    Cheese Dressing*

Snack
fresh apricots

Dinner
Vegetable Stirfry*
rice
fresh pear

DAY 24

Breakfast
long-cooking oatmeal
fresh grapefruit
coffee or tea

Lunch
lima beans
Barley with Mushrooms*
steamed spinach drizzled with
    olive oil

DAY 24 (continued)
Snack
Sweet Potato Fries*

Dinner
Black Bean Chili*
Rice Salad with Vegetables and
    Herbs*
green salad with Garlic and
    Buttermilk Dressing*

DAY 25

Breakfast
All-Bran cereal (Kellogg's)
low-fat milk
fresh apricots
coffee or tea

Lunch
roast lamb
baked acorn squash
green salad with tomatoes and
    Low-Fat Blue Cheese Dressing*

Snack
low-fat cottage cheese
celery sticks

Dinner
Spicy Bean Dip* sandwich with
    cucumber, sprouts, and tomato
2 slices Good Calorie Cracked
    Wheat Bread*
fresh orange

DAY 26

Breakfast
baked yam with butter
fresh grapefruit
coffee or tea

Lunch
broiled cod fillet with lemon
rice
steamed asparagus
low-fat chocolate ice cream

Snack
fresh cherries

Dinner
split pea soup
1 slice Good Calorie Cracked
    Wheat Bread*
green salad with Garlic and
    Buttermilk Dressing*

DAY 27

Breakfast
2 Good Calorie Bran Muffins*
Apple Butter*
coffee or tea

Lunch
Escarole and Chickpeas*
Yam and Cheese Strudel*
bulgur

Snack
yogurt-and-fruit shake

Dinner
Quick Egg Drop Soup*
Mashed Sweet Potatoes*

DAY 28

Breakfast
long-cooking oatmeal
fresh blueberries
coffee or tea

Lunch
artichoke with mayonnaise
Vegetable-Saffron Paella*
unsweetened applesauce

Snack
Hummus*
raw vegetables

Dinner
Summer Squash Soufflé*
barley
spinach salad with Low-Fat Blue
     Cheese Dressing*

HAVE NO FEAR

Finally, we urge you to relax and enjoy yourself. No longer will dieting attack your serotonin and turn you into a physical and mental wreck. Instead, you will be relieved to find you experience none of the problems that previously plagued your efforts to lose weight.

I bought your book, followed it completely, and now have achieved wonderful blood sugar readings. I have also lost that ten extra pounds I could never get off. I have never eaten so much good food, felt so wonderful and best of all, I never ever walk around hungry anymore.

—*R. A., Golden Valley, MN*

I went on a low-fat diet and vigorous exercise regime. Not only did I not lose weight, I gained ten more pounds. My groom, who never thought low-fat dieting was a good idea, picked up your book for me. As soon as I began following your diet plan I began to lose weight. Your research has really paid off, I once again enjoy a slim figure.

—*L. D., Denver, CO*

CHAPTER 8

Caloric Restriction

Why are we giving you a program of caloric restriction after we have already seen how conventional caloric-restriction diets induce low serotonin?

One of the signals your body uses to create its response to the threat of starvation is excess insulin and low serotonin. Good Calories prevent both of these signals from developing.

We suggest that you follow the caloric-restriction program if you need to lose less than fifteen pounds, if you have normal insulin, or if your excess weight is concentrated in your legs.

Remember, eating serotonin-stimulating foods will transform caloric restriction and turn it into a program that is easy to follow because you no longer suffer from the food cravings, depression, and irritability of low serotonin.

Additionally, you will lose weight faster than you would with conventional caloric restriction. The clinical study that proved the effectiveness of Good Calories also showed that they increased the rate of weight loss.

Caloric restriction is frequently needed to lose those last few pounds. Those last few pounds are generally the hardest to go but one thing is certain, you don't want to destroy your wonderful state of harmony and equilibrium to achieve it. Make sure to be very careful about eating Good Calories, otherwise you may find your serotonin chemistry is thrown off and so is your current state of equilibrium. Losing a few pounds is just too high a price to pay for that.

IF YOU DON'T LOSE WEIGHT ON THE FREE-EATING PLAN

If you do not lose weight on the free-eating program, you must ask yourself if you are following the program faithfully. The following are the most common ways that dieters deprive themselves of the benefits of the unrestricted program.

- not eating a *large* breakfast of *only* low-protein Good Calorie carbohydrates
- eating too much animal protein
- combining protein or saturated fats with starches
- snacking on Bad Calories
- skipping meals, especially after cheating
- eating too much saturated fat
- eating too little unsaturated fat
- eating too late at night
- not taking the supplements, particularly chromium

If you violated any of these rules you should start the free-eating program again. You never gave the program a chance to work.

If you have not violated these rules and are not losing weight it means you should consider the caloric-restriction program. It will be totally different from your image of dieting. Even caloric restriction is made easier by Good Calories.

CALORIC-RESTRICTION PLAN— 1,200 CALORIES A DAY

This is the same program as given in the free-eating chapter except that calories count and portions are detailed.

Remember that any eating plan should follow the six rules of a woman-friendly nutritional environment.

SIX PATHS TO A WOMAN-FRIENDLY ENVIRONMENT

1. Never skip meals.
2. Instant foods, processed foods, overly ripe foods, overcooked foods, and saturated fats should be avoided.
3. Overall, 50 to 60 percent of calories should be from carbohydrates, 20 percent from animal or vegetable protein, and 20 to 30 percent from fat.
4. Eat only Good Calorie carbohydrates for breakfast, eat a full breakfast, and eat little or no protein for breakfast.
5. Take vitamin and mineral supplements.
6. Substitute Good Calories for Bad Calories, and cheat intelligently.

Day	Serving Size	Calories
DAY 1		
Breakfast		
long-cooking oatmeal	1 cup	130
Good Calorie Bran Muffin*	1 muffin	109
fresh orange	1	62
black coffee or tea		0
Total		301
Lunch		
Pasta with Asparagus and Roasted Peppers*	1 serving	338
green salad with Low-Fat	unlimited greens	
Blue Cheese Dressing*	+ 4 Tbs. dressing	80
fresh apple	1	85
Total		503
Snack		
plain nonfat yogurt sweetened		
with 1 tsp. table sugar or		
artificial sweetener equivalent	1 cup	122
Dinner		
Lentil and Barley Pilaf*	1 serving	190
steamed broccoli with Parmesan cheese	1½ cups broccoli	
	+ 2 Tbs. Parmesan	
	cheese	72
Total		262

Day	Serving Size	Calories
DAY 2		
Breakfast		
Good Calorie Applesauce Bread*	2 slices	220
Apple Butter*	2 Tbs.	50
orange juice	6 oz.	83
black coffee or tea		0
Total		353
Lunch		
Roasted Vegetable and Pasta Salad*	1 serving	160
Spinach and Red Onion Stirfry*	1 serving	90
green salad with olive oil and	unlimited greens	
lemon dressing	+ 2 tsp. olive oil	80
Total		330
Snack		
apple juice	6 oz.	87
low-fat cottage cheese	½ cup	101
Total		188
Dinner		
Yellow Split Pea Soup*	1 serving	232
Rice Salad with Vegetables and Herbs*	1 serving	105
fresh peach	1	37
Total		374
DAY 3		
Breakfast		
baked yam with butter	1 yam, 1 pat of butter	216
fresh tangerine	1	37
black coffee or tea		0
Total		253
Lunch		
Creamy Rice and Beans*	1 serving	201
Steamed spaghetti squash with	1 cup +	
Parmesan cheese	1 Tbs. cheese	73
steamed artichoke	1	60
Total		334
Snack		
Good Calorie Oat Muffin*	1 muffin	107
fresh cherries	30	150
Total		257

Day	Serving Size	Calories
DAY 3 *(continued)*		
Dinner		
Asparagus Soup*	2 servings	162
Italian Frittata*	1 serving	149
spinach salad with Garlic and	unlimited spinach	
Buttermilk Dressing*	+ 4 Tbs. dressing	80
Total		391
DAY 4		
Breakfast		
Good Calorie Apricot–Oat Bran Muffins*	2 muffins	220
fresh apple	1	80
black coffee or tea		0
Total		300
Lunch		
lentil soup	1½ cups	370
green salad with olive oil and vinegar	unlimited salad	
	+ 1 tsp. olive oil	120
Total		490
Snack		
fresh fruit salad	1½ cups	150
Dinner		
Vegetable Stirfry*	1 serving	151
rice	1 cup	164
Total		315
DAY 5		
Breakfast		
long-cooking oatmeal	1½ cups	195
fresh blueberries	½ cup	41
black coffee or tea		0
Total		236
Lunch		
Wheat Berry Soup*	1 serving	257
Tabouli*	1 cup	136
green salad with Cottage Cheese Dressing*	unlimited greens	
	+ 4 Tbs. dressing	80
Total		473

Day	Serving Size	Calories
Snack		
Dry-roasted Curried Chickpeas*	½ cup	134
Dinner		
Spinach Lasagna*	1 serving	236
steamed green beans with olive oil	1½ cups green beans	
	+ 1 tsp. olive oil	91
fresh peach	1	37
Total		364

DAY 6

Breakfast		
All-Bran cereal (Kellogg's)	1 cup	210
low-fat milk	½ cup	61
black coffee or tea		0
Total		271
Lunch		
Vegetable Terrine*	1 serving	220
brown rice	½ cup	110
Asian pear	1	75
Total		405
Snack		
low-fat cottage cheese	½ cup	101
celery sticks	4	28
Total		129
Dinner		
pasta topped with olive oil	1 cup pasta	
and Parmesan cheese	+ 2 tsp. olive oil	
	+ 1 Tbs. Parmesan	286
steamed peas	1 cup	50
spinach salad with Low-Fat Blue	unlimited spinach	
Cheese Dressing*	+ 4 Tbs. dressing	80
Total		416

DAY 7

Breakfast		
Good Calorie Cracked Wheat Bread*	2 slices	230
butter	2 pats	72
black coffee or tea		0
		302

Day	Serving Size	Calories
DAY 7 *(continued)*		
Lunch		
plain pinto beans	½ cup	117
rice	½ cup	82
salsa	¼ cup	20
corn tortillas	2 tortillas	60
Total		279
Snack		
Eggplant Dip*	¼ cup	88
raw vegetables	1 cup	30
Total		118
Dinner		
Fresh Basil and Tomato Pasta*	1 serving	351
green salad with Yogurt Dressing*	unlimited greens	
	+ 4 Tbs. dressing	80
fresh apple	1	80
Total		511
DAY 8		
Breakfast		
long-cooking oatmeal	1 cup	130
fresh strawberries	1 cup	46
orange juice	6 oz.	83
black coffee or tea		0
Total		259
Lunch		
grilled chicken breast	5 oz.	269
string beans	1½ cups	51
green salad with olive oil and feta cheese	unlimited greens	
	+ 2 tsp. olive oil +	
	½ oz. feta cheese	165
Total		485
Snack		
fresh apple	1	80
Dinner		
Gazpacho*	1 cup	115
Pad Thai*	1 serving	281
Total		396

Day	Serving Size	Calories
DAY 9		
Breakfast		
German hard rye bread	2 slices	220
Apple Butter*	2 Tbs.	50
fresh grapefruit	½	46
black coffee or tea		0
Total		316
Lunch		
Steamed Flounder with Herbs*	1 serving	178
Margrit's Curried Brown Rice*	½ cup	195
green salad with	unlimited salad	
Garlic and Buttermilk Dressing*	+ 4 Tbs. dressing	80
Total		453
Snack		
Spicy Bean Dip*	½ cup	196
raw vegetables	1 cup	30
Total		226
Dinner		
Asparagus Soup*	1 serving	81
Roasted Vegetable and Pasta Salad*	1 serving	160
Total		241
DAY 10		
Breakfast		
baked yam with butter	1 yam, 1 pat of butter	216
fresh peach	1	37
black coffee or tea		0
Total		253
Lunch		
Lentil and Spinach Stew*	1 serving	165
turkey sandwich with lettuce, tomato,	3 oz. turkey,	
and mustard on Good Calorie	2 slices	
Sauerkraut Rye Bread*	bread	350
fresh strawberries	½ cup	23
Total		538
Snack		
fresh strawberries	½ cup	23
Total		23

Day	Serving Size	Calories
DAY 10 (continued)		
Dinner		
Fresh Basil and Tomato Pasta*	1 serving	351
steamed zucchini	1½ cups	45
Total		396
DAY 11		
Breakfast		
Good Calorie Bran Muffins*	2 muffins	218
fresh grapefruit	½	46
black coffee or tea		0
Total		264
Lunch		
Shrimp and Pea Pod Stirfry*	1 serving	410
rice	½ cup	82
Total		492
Snack		
fresh apricots	3	57
Dinner		
Black Bean Chili*	1 serving	285
green salad with	unlimited greens	
Garlic and Buttermilk Dressing*	+ 4 Tbs. dressing	80
Total		365
DAY 12		
Breakfast		
All-Bran cereal (Kellogg's)	1 cup	210
low-fat milk	½ cup	61
fresh strawberries	½ cup	23
black coffee or tea		0
Total		294
Lunch		
lamb steak	4 oz.	250
steamed broccoli with lemon	1½ cups	36
green salad with Creamy Thousand	unlimited greens	
Island Dressing*	+ 3 Tbs. dressing	82
fresh figs	3	120
Total		488

Day	Serving Size	Calories
Snack		
Sweet Potato Fries*	1 serving	115
Dinner		
Lentil and Barley Pilaf*	1 serving	190
cucumber salad with vinaigrette	½ cucumber	
	+ 2 tsp. oil	99
Total		289

DAY 13

Breakfast		
long-cooking oatmeal	1 cup	130
firm banana	1	100
black coffee or tea		0
Total		230
Lunch		
Black Bean and Crab Bisque*	1 serving	290
Vegetable–Cottage Cheese Salad*	1 serving	97
Good Calorie Sauerkraut-Rye Bread*	1 slice	115
Total		502
Snack		
Eggplant Dip*	¼ cup	88
raw vegetables	1 cup	30
Total		118
Dinner		
Poached Fish with White Wine and Fennel*	1 serving	254
Summer Squash Soufflé*	1 serving	92
green salad with Yogurt Dressing*	unlimited greens	
	+ 4 Tbs. dressing	64
Total		410

DAY 14

Breakfast		
Good Calorie Oat Bread*	2 slices	264
fresh peach slices	1 cup	54
black coffee or tea		0
Total		318

Day	Serving Size	Calories

DAY 14 *(continued)*
Lunch

pasta	1½ cups pasta	260
tomato sauce	½ cup bottled sauce	50
Sesame Green Bean Salad*	1 serving	<u>61</u>
Total		371

Snack

Dry-roasted Chickpeas*	½ cup	134

Dinner

steamed vegetables	2 cups	60
low-fat tofu	½ cup	94
Tamari-Ginger Sauce*	¼ cup	17
rice	½ cup	82
Cucumber, Tomato, and Yogurt Salad*	1 serving	<u>111</u>
Total		364

DAY 15

Breakfast

Good Calorie Cracked Wheat Bread*	2 slices	230
unsweetened strawberry jam	2 tsp.	25
orange juice	6 oz.	63
black coffee or tea		<u>0</u>
Total		318

Lunch

Steamed Flounder with Herbs*	1 serving	178
new potatoes	3	120
green salad with	unlimited greens	
Cottage Cheese Dressing*	+ 4 Tbs. dressing	<u>80</u>
Total		378

Snack

Curried Yogurt Dip*	½ cup	44
raw vegetables	1 cup	<u>30</u>
Total		74

Dinner

Quick Garlic Soup*	unlimited	6
Middle Eastern Lentils and Rice*	1 serving	<u>449</u>
Total		455

Day	Serving Size	Calories
DAY 16		
Breakfast		
long-cooking oatmeal	1½ cups	195
fresh nectarine	1	67
black coffee or tea		0
Total		262
Lunch		
tomato soup	1 cup	75
grilled coho salmon	5 oz.	263
steamed kale	1½ cups	63
Total		401
Snack		
hot carob drink	8 oz. low-fat milk + 3 tsp. unsweetened carob powder	166
Dinner		
steamed vegetables	2 cups	60
barley	1 cup	193
Tamari-Ginger Sauce*	½ cup	17
unsweetened applesauce	1 cup	106
Total		376
DAY 17		
Breakfast		
long-cooking oatmeal	1½ cups	195
fresh raspberries	1 cup	82
black coffee or tea		0
Total		277
Lunch		
grilled chicken breast	5 oz.	269
steamed zucchini and yellow squash	2 cups	36
green salad with Creamy Thousand Island Dressing*	unlimited greens + 3 Tbs. dressing	80
Total		385
Snack		
Sweet Potato Fries*	1 cup	115

Day	Serving Size	Calories
DAY 17 *(continued)*		
Dinner		
Quick Egg Drop Soup*	1	70
Spinach Fettuccine with		
Sun-dried Tomatoes*	1 serving	336
steamed asparagus	1½ cups	66
Total		472
DAY 18		
Breakfast		
steamed yam with butter	1 yam,	
	1 pat of butter	216
fresh grapes		114
black coffee or tea		0
Total		330
Lunch		
Shrimp with Spicy Black Bean Sauce*	1 serving	297
rice	½ cup	82
fresh apple	1	80
Total		459
Snack		
Curried Yogurt Dip*	½ cup	44
raw vegetables	1 cup	30
Total		74
Dinner		
Quick Vegetable Soup*	1 serving	65
Spinach Lasagna*	1 serving	236
green salad with Garlic and Buttermilk	unlimited greens	
Dressing*	+ 4 Tbs. dressing	80
Total		381
DAY 19		
Breakfast		
long-cooking oatmeal	2 cups	260
orange juice	6 oz.	63
black coffee or tea		0
Total		323

Day	Serving Size	Calories
Lunch		
Lemon Chicken*	1 serving	165
steamed broccoli	1½ cups	36
cucumber salad with	½ cucumber	
olive oil vinaigrette	+ 2 tsp. olive oil	_58_
Total		259
Snack		
Good Calorie Oat Bread*	1 slice	132
low-fat cottage cheese	½ cup	_101_
Total		233
Dinner		
Chickpea Salad with		
Tomatoes and Vinaigrette*	1 serving	222
steamed artichoke	1 medium	60
fresh apple	1	_80_
Total		362

DAY 20

Day	Serving Size	Calories
Breakfast		
Good Calorie Cracked Wheat Bread*	2 slices	230
Apple Butter*	2 Tbs.	50
black coffee or tea		_0_
Total		280
Lunch		
broiled halibut	5 oz.	199
steamed cauliflower	1½ cups	45
broiled tomato with	1 tomato +	
Parmesan cheese	2 Tbs. Parmesan	76
spinach salad with Yogurt	unlimited spinach	
Dressing*	+ 4 Tbs. dressing	80
fresh plum	1	_36_
Total		436
Snack		
Eggplant Dip*	¼ cup	88
raw vegetables	1 cup	_30_
Total		118

Day	Serving Size	Calories

DAY 20 (*continued*)
Dinner

steamed vegetables	2 cups	60
low-fat tofu	½ cup	94
Tamari-Ginger Sauce*	½ cup	35
barley	1 cup	<u>200</u>
Total		389

DAY 21

Breakfast

All-Bran cereal (Kellogg's)	1 cup	210
low-fat milk	½ cup	61
fresh strawberries	½ cup	23
black coffee or tea		<u>0</u>
Total		294

Lunch

Good Calorie Sauerkraut-Rye Bread*	1 slice	115
Black Bean and Crab Bisque*	1 serving	290
green salad with	unlimited greens	
olive oil vinaigrette	+ 2 tsp. olive oil	<u>80</u>
Total		485

Snack

fresh plum	1	37

Dinner

Lemon-Bulgur Timbales*	1 serving	150
Sumptuous Broccoli*	1 serving	122
baked yam with salsa	1 yam, ¼ cup salsa	<u>180</u>
Total		452

DAY 22

Breakfast

Good Calorie Apricot–		
Oat Bran Muffins*	2	220
unsweetened strawberry jam	2 tsp.	25
black coffee or tea		<u>0</u>
Total		245

Day	Serving Size	Calories
Lunch		
Elegant Fish Pâté*	1 serving	311
steamed asparagus	1½ cups	60
green salad with Creamy Thousand	unlimited greens	
Island Dressing*	+ 3 Tbs. dressing	80
Total		451
Snack		
fresh blueberries	½ cup	41
vanilla low-fat ice cream	1 cup	120
Total		161
Dinner		
Pasta with Fresh Vegetables*	1 serving	354

DAY 23

Breakfast		
Good Calorie Applesauce Bread*	2 slices	220
fresh grapes	1 cup	60
black coffee or tea		0
Total		280
Lunch		
Hummus* sandwich with lettuce,		
tomato, and sprouts	½ cup	200
Good Calorie Cracked Wheat Bread*	2 slices	230
green salad with Low-Fat	unlimited greens	
Blue Cheese Dressing*	+ 3 Tbs. dressing	79
Total		509
Snack		
fresh apricots	3	57
Dinner		
Vegetable Stirfry*	1 serving	151
rice	1 cup	160
fresh pear	1	98
Total		409

DAY 24

Breakfast		
long-cooking oatmeal	1½ cups	195
fresh grapefruit	½	46
black coffee or tea		0
Total		241

Day	Serving Size	Calories

DAY 24 *(continued)*
Lunch

lima beans	½ cup	136
Barley with Mushrooms*	1 serving	145
steamed spinach drizzled with olive oil	1½ cups + 2 tsp. oil	<u>143</u>
Total		424

Snack

Sweet Potato Fries*	1 serving	115

Dinner

Black Bean Chili*	1 serving	285
Rice Salad with Vegetables and Herbs*	1 serving	105
green salad with Garlic and	unlimited salad	
Buttermilk Dressing*	+ 4 Tbs. dressing	<u>80</u>
Total		470

DAY 25

Breakfast

All-Bran cereal (Kellogg's)	1 cup	210
low-fat milk	½ cup	61
fresh apricots	3	57
coffee or tea		<u>0</u>
Total		328

Lunch

roast lamb	4 oz.	231
baked acorn squash	1	100
green salad with tomatoes and	unlimited salad	
Low-Fat Blue Cheese Dressing*	+ 3 Tbs. dressing	<u>80</u>
Total		411

Snack

low-fat cottage cheese	½ cup	101
celery sticks	7	<u>28</u>
Total		129

Dinner

Spicy Bean Dip* sandwich with cucumber slices, sprouts, and tomato	¼ cup	98
Good Calorie Cracked Wheat Bread*	2 slices	230
fresh orange	1	<u>65</u>
Total		393

Day	Serving Size	Calories

DAY 26

Breakfast

baked yam with butter	1 yam,	
	1 pat of butter	216
fresh grapefruit	½	46
black coffee or tea		0
Total		262

Lunch

broiled cod fillet with lemon	6 oz.	179
rice	½ cup	82
steamed asparagus	1½ cups	66
low-fat chocolate ice cream		120
Total		447

Snack

fresh cherries	20	98

Dinner

split pea soup	1½ cups	240
Good Calorie Cracked Wheat Bread*	1 slice	115
green salad with Garlic and	unlimited greens	
Buttermilk Dressing*	+ 4 Tbs. dressing	80
Total		435

DAY 27

Breakfast

Good Calorie Bran Muffins*	2 muffins	218
Apple Butter*	2 Tbs.	50
black coffee or tea		0
Total		268

Lunch

Escarole and Chickpeas*	1 serving	118
Yam and Cheese Strudel*	1 serving	150
bulgur	1 cup	152
Total		420

Snack

yogurt-and-fruit shake	1 cup nonfat yogurt	
	+ ½ cup fruit +	
	1 tsp. table sugar or	
	artificial sweetener	
	equivalent	150

Day	Serving Size	Calories

DAY 27 *(continued)*
Dinner

Quick Egg Drop Soup*	1 serving	70
Mashed Sweet Potatoes*	1 serving	<u>278</u>
Total		348

DAY 28

Breakfast

long-cooking oatmeal	1½ cups	195
fresh blueberries	1 cup	82
black coffee or tea		<u>0</u>
Total		277

Lunch

artichoke with mayonnaise	1 artichoke + 1 tsp. mayonnaise	93
Vegetable-Saffron Paella*	1 serving	279
unsweetened applesauce	1 cup	<u>100</u>
Total		472

Snack

Hummus*	¼ cup	100
raw vegetables	1 cup	<u>30</u>
Total		130

Dinner

Summer Squash Soufflé*	1 serving	92
barley	⅔ cup	129
spinach salad with Low-Fat Blue Cheese Dressing*	unlimited spinach + 4 Tbs. dressing	<u>80</u>
Total		301

CHAPTER 9

Decreasing Stress

The good news is that a woman's naturally high levels of serotonin allow her to withstand stress better than a man.

The bad news is that a woman under stress has a greater need for serotonin to replace the serotonin she uses to fight stress. This need is so great that her blood becomes depleted of the amino acid tryptophan, which is made into serotonin.[1] A woman's ability to make new serotonin is even further depleted if she is under stress and on a diet.

Stress burns away more of a woman's serotonin than a man's. Studies have shown that repeated stress reduces the serotonin of female, but not male, rats.[2] Female rats use more serotonin when they are subjected to inescapable electrical shocks.[3] The stress of separating a child from its mother attacks a female monkey's serotonin more than a male's serotonin.[4]

Stress also decreases a woman's ability to make estrogen. Not only will this interfere with her reproductive cycle, but it also inhibits her ability to make serotonin.

When a woman can no longer make enough serotonin, her response to stress is the same as a man's.[5] Her inherent advantage has been diminished.

Stress also increases appetite and insulin, creating food addiction. Mental stress alters the sympathetic nervous system, which helps determine appetite and food preferences, releasing hormones that exert direct effects on blood-sugar metabolism.[6] Stress acts through the glucocorticoid system to promote the excess insulin associated with insulin resistance.[7] Stressed women also excrete essential vitamins and minerals that should be retained if insulin levels are to remain under control.

Because their insulin levels are out of control, stressed women are more susceptible to the food addiction that accompanies insulin surges.

The simple act of dieting floods a woman's body with stress-producing hormones. Prolactin is a hormone released during mental or physical stress. After dieting, a woman's prolactin rises by more than 50 percent while a man's prolactin remains unchanged.[8]

Adding it all up, we see that dieting creates stress in women and makes it nearly impossible for them to make enough serotonin to withstand that stress. At the same time, it creates conditions that can lead to food addiction.

Clearly, dieting women have a special need to reduce stress in their lives. This chapter introduces three techniques for lowering stress: exercise, meditation, and breathing.

We all know that reducing stress boosts your physical and mental health. Now it's time to see that it's also a powerful tool to make weight loss easier and to protect a woman's source of personal power.

EXERCISE

Most weight-loss programs recommend exercise as a means to burn excess calories. Unfortunately, many of these programs can be excessive for some women. Instead, we are going to focus on what more balanced, gentler forms of exercise can do for a woman's mind and for her levels of serotonin and insulin.

The right kind of exercise assists any weight-loss program. Not only does it relieve stress and help prevent depression, it also corrects imbalances in blood sugar and insulin.[9] As an added bonus, it helps inhibit fat formation.[10]

Some research suggests that exercise elevates serotonin.[11] However, it is equally possible that exercise functions mainly to prevent stress from lowering serotonin.[12] In either case, exercise is a powerful addition to any woman's program.

However, you shouldn't overdo exercise. Unfortunately, many dieting women attack their bodies with such a fury of exercise that prolactin, serotonin, and other important hormones go into the stress mode. Running a marathon actually lowers tryptophan entry into the brain.

The key is to be gentle with your body. All you need is about thirty minutes of exercise per day. You can do more, but only if you gradually build up to a higher level.

The "runner's high" that follows all forms of extreme exercise has been wrongly identified as a surge in serotonin. Instead, this high is a symptom of your body releasing natural opiates called endorphins to relieve the pain and stress brought on by overexercising. Bringing on more stress is the last thing you want to do. Exercise in moderation—do not overexercise.

Our program does not recommend high-intensity exercise. Instead, we suggest you follow these guidelines:

1. Find an exercise you enjoy.
2. Exercise for at least thirty minutes, five times per week.
3. Do not overdo it. Never exercise to exhaustion.

What form of exercise should you adopt? Some authorities suggest that repetitive muscle movements are most effective at elevating serotonin. This would suggest that aerobics, walking, and running are effective forms of exercise.

Hatha yoga and tai chi are also excellent forms of exercise. These ancient disciplines are based on a balanced approach to fitness. They incorporate the whole person—the entire mind-body connection. These forms of exercise make you feel energized, calm, and revitalized. We recommend them highly.

Any form of exercise won't do you any good, however, if you don't do it. Ultimately, the best form of exercise is one that you'll do. Walking, gardening, raking leaves, playing golf, dancing—do anything that gets you moving!

MEDITATION

Meditation is the ultimate way to reduce stress. Meditation seems to work on all the markers of stress, lowering heart rate and blood pressure.[13] Meditation also reduces the hormones associated with stress, prolactin and glucocorticoids.[14]

Meditation also counters a dieting woman's tendency toward depression and low serotonin. Meditation lowers the thyroid-stimulating hormone (TSH), which creates depression.[15] Meditation also increases serotonin activity while decreasing the activity of the neurotransmitters that oppose serotonin.[16]

One thing is for certain: If you meditate you will discover the stresses of daily life are not as great as you previously imagined.

What is meditation? It is like concentration, something we already do all the time. Concentration is necessary to accomplish anything in life. As you read our words, you are concentrating. If you don't concentrate when you're driving, your car wanders all over the road. If you don't concentrate when reading, the words don't make sense. If you don't concentrate on your work, you're not productive. Concentration is the continual flow of our attention on an object or activity.

The difference between concentration and meditation is the direction of focus. Whereas concentration is externally directed, meditation is directed to our inner self, the deepest part of our being. This is where we experience our freedom and perfection. Meditation is a means to experience who we really are. Stress and tension are actually expressions of who we are not. When we focus our attention inward through meditation, our true qualities begin to radiate—happiness, freedom, beauty, and peace. Experiencing great joy is the greatest way we know of to obtain relief from stress!

Proper breathing and a mantra are tools to help us experience meditation. An important thing to realize about meditation is that it doesn't occur only while sitting in a certain posture. A meditative experience can happen anytime, anywhere. When you're feeling stressed you can repeat the mantra and do the following breathing exercises to experience a meditative feeling. It can be done quietly and inconspicuously. No one has to know. Freedom from stress will follow, inevitably, every time, even if you are out and about. Remember that meditation is simply the flow of our attention toward our inner self. Your inner self never leaves you; it accompanies you everywhere, whether you are sitting in meditation, cooking dinner, or negotiating in a boardroom.

A mantra is a phrase to help you enter meditation more easily.

Eventually, the connection between meditation and mantra becomes so strong that simply repeating the mantra silently lets you immediately enter into a more meditative awareness.

We recommend the "Hamsa" mantra. Although it is very simple, it is also extremely powerful. It's as simple as listening to your breath. If you listen closely to your breath as it enters and leaves your body you can hear its subtle sound, Hamsa. "Ham" (pronounced *hamm*) is the sound incoming breath makes. "Sa" (pronounced *saahh*) is the sound of the outgoing breath. As you focus on the sound of the breath, and either listen to the sound of Hamsa or repeat the words to yourself, your breath becomes quieter and steadier. It might help to know that Hamsa is the actual vibrational sound of our inner self. As we listen to it, we are remembering and honoring the truth within us. (Please note: This mantra is universal; it does not have any particular religious affiliation. People from all religious persuasions use it to connect with their inner selves.)

Observing your breath in this way leads to relaxation. In fact, as you change your breathing pattern you change your emotional state. Have you ever noticed what happens to your breathing when you're afraid or angry? When you're angry you alternate between holding your breath and taking deep gulps of air. When you are nervous or afraid, your breath will be rapid and shallow. Or, when you sit down to relax after exerting yourself, you take a long, deep breath. When you are calm, your breathing is even, deep, and slow. Your breath mirrors your thoughts and emotions. A simple way to change the way we feel is to change how we breathe. If you don't want to use the Hamsa mantra, you can simply focus on your breath. Just focusing on the breath can give you an enormous boost of energy and mental clarity. You will find as you pay attention to your breath, it becomes quieter and you become more peaceful and centered.

The way to breathe during meditation is the way you naturally breathe when you are calm: slowly, evenly, and deeply. Breathing this way makes your mind receptive to meditation.

Meditation Instructions

Breathe deeply.

Coordinate the mantra with the rhythm of your breathing.

As you inhale, silently repeat "ham."

As you exhale, silently repeat "sa."

Hear these sounds in your breath.

Put the book down for a moment and try this technique.

Now you are ready to start regular meditation sessions.

It is best if you incorporate certain constants. The more constants there are in any situation the more quickly the mind recognizes and responds to it.

It helps to meditate at approximately the same time each day. Analyze your daily routine. Look for moments when some slack time occurs. Look for times when you are most likely to need rejuvenation and clarity. Also, early mornings and late evenings before going to bed are excellent times because the environment is generally quieter. Whatever time you choose, try to be regular. Regularity of the sessions helps the mind anticipate what's coming and respond more quickly. For the same reason, it is helpful to sit in the same place every day. Ideally, you should meditate in a quiet room in which you can be alone. If you don't have that luxury, just find a quiet corner and make it your meditative haven. Use earplugs if you can't find a quiet spot.

Begin with fifteen to twenty minutes. Sit comfortably, either on the floor with your legs crossed or in a chair. Maintain a relaxed, comfortable, upright position. Your back and neck should be straight to allow for full, deep breathing. But you don't want to spend your session focusing on physical pain. Keep your back straight without stiffening your muscles. Lean against a wall if necessary. Sometimes cushions under the buttocks or knees help the back relax more. Turn off the phone and remove any other possible distraction that might grab your attention away from meditation. It is helpful to set a timer to signal the end of the session so you're not constantly looking at your clock.

What can you expect when you sit down to meditate?

One thing you can count on is that your mind will wander. Nothing is more natural. Your mind is constantly busy creating thoughts and images. Most often first-time meditators are shocked to see how busy their minds are. This is very good to see. Most of the time we don't even notice what's going on in our heads; it's happening automatically. Already you've become more aware! Use the breath and mantra to keep bringing you to a place deeper than your thoughts.

Often the mind will rebel against this activity. Don't worry. The mind causes *everyone* problems. Your mind will remind you of all the important things you must be doing—what foods you should be preparing, the letters that need to be written, or the laundry that must be done. You'll wonder why you're even doing this; after all, you have so little time. And on and on. This is a good time to remind yourself that sitting "doing nothing" is actually a beneficial and productive thing. When the benefits of meditation begin to manifest, it will be a lot easier to be convinced of this, but in the beginning you'll just have to trust that this is so!

During meditation you may experience spontaneous images or sounds. It's very normal. Just observe them and enjoy the feelings that come with them. Just let your innate intelligence reveal itself to you and watch everything in the same spirit that you would watch a movie, with detached interest. When you are watching a movie you allow the story to unfold without getting all worked up about every passing image. You know it's a movie and you just enjoy the experience. Have the same attitude in meditation. When a thought or image arises just witness and observe it. Don't engage it; just let it be.

If you find yourself running off with particular thoughts, just bring your attention back to your breath and the mantra, Hamsa.

Gradually you will find yourself being less controlled by the thoughts in your mind. Meditation will give you the ability to experience the calm that lies beyond the mind. Even though thoughts will go on, you'll be less bothered by them, and you will be able to go into meditation easily. This will happen not only when you're sitting for formal meditation; it will also occur when you are engaged in your normal daily routine. Just mentally repeat the mantra with your breath. Your mind will go where it is accustomed to going when its hears the mantra: to inner calm.

ADVANCED EXERCISES: BREATH RETENTION

For readers who already meditate and want a more advanced breathing technique, we have included other Hamsa breath practices. If you are a new meditator you should not attempt these until you feel comfortable with your basic meditation.

Practice

Repeat the Hamsa meditation as previously described. Synchronize the syllable "ham" with the incoming breath and "sa" with the outgoing breath. The moment between the in and out breath is a moment of consummate peace. Focus on the peace of these moments of suspension. Notice how perfectly calm and still these moments are.

Now, extend the duration of these moments of peace. Allow the time between inhalation and exhalation to extend.

Retain your breath for increasing periods of time. Keep your breath even, don't aggressively try to maintain the inhalation or exhalation. The breath should be steady and calm. Gradually you will be able to increase the duration of breath retention each time you perform the exercise. As the time between inhalation and exhalation increases, your peace will increase too.

Feel it happen.

Remind yourself that your peace is expanding.

Pay attention to how you feel afterward. Notice how the peace stays with you.

Advanced Practice

After you are comfortable with the basic practice, and only after you feel its benefits, you may want to try this advanced practice. Perform the breathing exercises as described above, focusing on the following imagery.

Imagine that a temporary mental blank arises when you retain your breath. Imagine this puts you into a state of perfect balance in which the secrets of the universe are revealed to you.

Imagine that, from your normal existence, you are entering a realm where even routine states of mind are richer and more powerful than you ever imagined possible.

Very Advanced Practice

Perform these exercises while going about your daily life. Your breath is always with you so there is no reason to separate "meditative" breathing from your usual routine. You will find you can meditate at will.

CHAPTER 10

Questions and Answers

The following are questions we have received from people who tried our programs.

Question: I've been following your diet for some time now with good results. I have lost thirty-two pounds and have plateaued. I currently eat whatever amount of food I like and stop when I am satisfied. I no longer have cravings and intense mood swings. I would like to lose another ten pounds, however. I am five feet, eight inches tall and weigh 135 pounds. Any suggestions for losing the last ten pounds?

Answer: Congratulations! Think of how wonderful your accomplishment is. You were able to lose thirty-two pounds eating until you were satisfied *and* your brain chemistry normalized to stop your cravings and mood swings.

You should know that your current weight is well within the healthy range for your height. The female ideal we have today is much thinner than ever before. At no time in history did we put a demand on women to be this thin. In fact, being thin used to signify either poor health or poverty.

However, that being said, if you are still interested in losing an additional ten pounds you'll probably have to opt for some caloric restriction. Try our 1,200-calories-a-day program *being sure to be very vigilant about eating predominantly Good Calories.* This is the only way to make caloric restriction work without damaging female serotonin brain chemistry. You'll know if your food cravings and/or mood swings return that you're either not eating enough or not being strict enough about eating mostly Good Calories.

Question: Does the free-eating plan mean that I can eat until I am full? How do I know how much to eat?

Answer: Your body will know; just give it a little time. As the brain's serotonin chemistry is balanced the body's innate intelligence can take over. After the first few days to one week of following the plan, your food cravings will diminish if you are one of the 40 percent who can use this program. You'll be less drawn to eating high-sugar and high-fat foods. Some women find they are eating less and yet are satisfied. Others, especially those who were eating low-fat and traditional low-calorie diets, find they can eat more and still maintain their optimal weight.

Question: Can you mix and match your recipes or do the menus you have suggested balance the Good and Bad Calories for the day?

Answer: The menus were designed to be a *guideline* for eating. It is highly unlikely that you could make all the different meals suggested unless you had a cook or very loving and devoted partner who has a lot of time. Most of us, however, might cook one recipe and use it for two meals. Or, make a pot of beans to use for a few days. This is perfectly acceptable. The menus are a guide of what kinds, combinations, and quantities of things to choose. Most of the menus are predominantly made up of Good Calories, however, to show you how easy and satisfying it is to eat this way.

Question: What about dairy foods? Although they produce low blood sugar, they also contain protein and fat. Are they Good or Bad Calories?

Answer: They present special problems for breakfast. Because protein can interfere with serotonin synthesis it is wise to avoid dairy products during breakfast. The occasional use of a small amount of milk in coffee or on cereal is permissible, but you should avoid breakfasts of yogurt or cheese.

Table 10-1 lists the glycemic indexes of some dairy foods. Notice that many dairy products vary according to how much added sugar or fruit is used in their manufacture.

TABLE 10-1
Dairy Foods
(Good Calories are in boldface.)

Ice cream	51–114
Milk	
Whole	**39**
Skim	**46**
Chocolate, sugar sweetened	**49**
Chocolate, artificially sweetened	**34**
Custard, milk	**61**
Nondairy frozen fruit product	**40**
Yogurt	
Low-fat, artificially sweetened	**20**
Low-fat, fruit, sugar-sweetened	47–110

Question: You mention that when a society eats a more vegetarian-based diet they are more likely to revere women. What about India? Women do not have much political or social power. Doesn't this conflict with your theory?

Answer: We also mention that during times of periodic starvation women's status in society declines. This can be when women subject themselves to low-calorie diets or when there is a general state of starvation in the land. India certainly fits into this latter case. However, in the Hindu religion women have long been regarded and honored as the embodiment of divine female energy, the goddess Shakti, so perhaps their predominantly vegetarian diet helps keep this ideal alive in the culture.

Question: I have been increasing my intake of high-fiber foods and legumes, which sometimes results in gas. What do you suggest?

Answer: It is not uncommon for people who switch to a diet with a higher fiber content to develop gas. This usually lasts for the first month as you are getting used to it. Some people find products like Beano help. (Beano is a liquid enzyme available in drugstores and health-food stores.) Also, soaking beans overnight, or for more than a day, before cooking them also helps. But stick with it; the benefits of eating a healthier, high-fiber diet are worth it. Also, try different kinds of beans. Philip finds that garbanzo beans produce the least gas for him.

Question: I have celiac sprue, a genetic disease whose only treatment is to eat a gluten-free diet. That eliminates my eating rye, wheat, barley, oats, bran, and pasta. I am asking for your help in planning my menus.

Answer: We were totally surprised to get this question because Philip also suffers from this condition and has lost thirty-five pounds on the program. So he is a living testament that it can work!

Rice and corn are the grains that celiacs must depend on. There are some delicious rice pastas that can be used in any of the recipes in this book. A special note: Because corn grain products and most products made with rice flour, such as rice pasta, are not Good Calories, it is better to eat them with something that is, such as legumes or olive oil, as this will lower their insulin-raising ability.

Question: Do you have any suggestions for quick meals?

Answer: There are plenty of quick and delicious meals you can prepare. Here are just a few.

- ◆ Pasta is always a great standby. Serve it with a few sautéed vegetables, a splash of olive oil, and grated Parmesan. If you want to give it more substance, add some beans.

♦ Make extra pasta (such as spirals or elbows) and use it to mix with greens, diced raw vegetables, and dressing for a hearty salad the next day.

♦ Use your favorite bottled tomato sauce over cooked pasta.

♦ Make a quick stirfry in a wok or large frying pan. Just use a little oil, ginger, garlic, vegetables, and tamari or soy sauce and stirfry, serving over rice. Or, for fried rice, add rice to the vegetable mixture, and add a beaten egg or egg white.

♦ Legumes (beans) freeze very well. Make extra bean soups, chilies, or cooked beans and freeze in freezer pouches. Microwave or reheat on the stove. Also, beans can be stored in glass jars for up to six months. So cook a little extra and bottle or freeze the beans to use when you don't have time. This will save you money as well as time.

Question: What about spices (cinnamon, nutmeg, etc.)? Good or bad?

Answer: All spices are fine. Use of salt, of course, should not be excessive. Sea salt has more flavor than regular salt so you need less of it.

Question: Are all nonfat dairy products okay?

Answer: Yes. However, the best time for a woman to eat dairy products is anytime but breakfast. Dairy products contain protein. The amount of protein a woman should eat at breakfast should be small to none. A little milk in your coffee or on cereal is fine occasionally, but save larger servings of dairy products for any other time in the day.

Question: Can you summarize the program so I can easily explain it to others?

Answer: The program breaks down to 50 percent carbohydrates, 20 to 25 percent protein (animal or vegetable), and 25 to 30 percent fat. The carbohydrates to choose from are those that raise serotonin gradually and do not produce insulin rushes. Those are seen on a glycemic index list with a value lower than 80. There is a strong emphasis on a good, healthy breakfast made up of the proper carbohydrates. It's that simple. Basically, it's a very healthy way to eat.

Question: I would like to get my teenage daughter on the program but she won't take a bag lunch with her to school. Is there anything I can do?

Answer: The most important thing is to send her off with a good, substantial breakfast. This is a great habit to get her into as soon as possible. Eating a breakfast of serotonin-enhancing Good Calories will help set her up for the entire day. Her mind will be clearer, her memory better, she'll feel more confident, and her sugar cravings will be reduced. She'll be less likely to binge out and will feel more in control. As she learns to eat Good Calories she'll also lose weight if she needs to. (This is an important consideration as we have a generation of the most obese children in history.)

Most importantly, get her away from the low-calorie-diet consciousness that has been ingrained in women of the last few generations. This will prevent her from suffering the detrimental effects that yo-yo dieting has on the mind and body. Tell her she can eat more of the right foods, and make some delicious meals for dinner to show her by example.

If you know that she is going to eat Bad Calories for lunch then you might want to serve Good Calorie dinners of principally carbohydrates and vegetable protein sources. This will counter her lunches and prime her for a serotonin-stimulating breakfast.

Question: Do you have any other quick breakfast suggestions?

Answer: Many cultures eat a breakfast that looks more like our lunch—for example, rice, broth, and vegetables. Try reheating some of yesterday's vegetarian lunch or dinner leftovers for a change.

Philip likes yams or rice with either spices or a small amount of sweetener. When he is not feeling well he makes a rice porridge by making rice with twice as much water as normal.

Monika likes to make long-cooking oatmeal, store the extra, and then microwave the rest in the morning.

Question: What can I use to sweeten my oatmeal in the morning?

Answer: Try adding a little fruit to the cooking oatmeal during the last minute or so—anything such as some chopped apple, blueberries, or strawberries, and sprinkle on some cinnamon. Or try a teaspoon of table sugar or a few drops of honey. These high–glycemic index sweeteners will not create a problem if they are added to Good Calories with low glycemic indexes; just be careful not to add too much sweetener.

Question: I see that carrots and potatoes are not on the Good Calorie list. Does this mean that I should never eat them?

Answer: No, of course not. Just learn the best way to eat them. Carrots and potatoes have a higher sugar content—just ask any diabetic! When eating carrots or potatoes, be sure not to overcook them. Eat them with foods that are on the Good Calorie list. For example, add half a chopped carrot to a stirfry of *other* vegetables. Or, eat a *firm* baked potato with a little olive oil, cottage cheese, or salsa instead of butter.

Question: My ten-year-old son was diagnosed with Attention Deficit Disorder. Some doctors think that ADD is a disorder affected by serotonin imbalances. Is it safe to put my son on this eating plan?

Answer: Absolutely. It's hard to criticize the eating plan we suggest. After all, it's based on eating a predominantly vegetarian diet, with lots of fruits and vegetables, legumes, whole unprocessed grains, pasta, and a moderate amount of fat. What's been eliminated are overprocessed foods, high-sugar carbohydrates, and excessive meat. It's certainly not a fad diet. Try adding these foods to his diet and see if he doesn't respond favorably.

It is interesting to note that all children, even boys, have the same serotonin chemistry that we describe in women, so they will be adversely affected in similar ways that women are by traditional diets. The reason for this is that until adolescence there is not a great enough difference in sex hormones to create a significant difference in the serotonin of boys and girls.

Question: Is it possible to continue on this program and lose all the weight I want without counting calories?

Answer: Many women, about 40 percent, will be able to. Some women who want to be really slender might have to use our caloric restriction program for those last ten to fifteen pounds. It depends on your unique characteristics. We recommend you stay on the unrestricted eating program until you aren't losing any more weight. Then, of course, reassess whether you really want or need to lose more. If you do, then go on the caloric-restriction program, being *absolutely* vigilant about eating Good Calories to prevent symptoms of low serotonin: depression, food cravings, irritability, inability to think clearly, and so on.

Question: Will I lose weight faster on the caloric-restriction plan?

Answer: Probably, but it is not recommended to lose more than two pounds per week, unless you are under a doctor's orders to lose weight more quickly. So go on the caloric-restriction plan only if you've stopped losing weight on the free-eating plan.

Question: I am not overweight. However, I do suffer from PMS. Could your program help me?

Answer: Recent scientific studies have determined that PMS is a serotonin affliction and serotonin-enhancing drugs like Prozac have proven to be effective in treating women with PMS. Because our program is based on serotonin-enhancing foods it should help. Obviously we can't make any claims regarding our program, but what harm can it do? Just eat lots of delicious, satisfying meals with Good Calories. It's certainly worth a try.

Question: With all this talk about how great serotonin is, I can't help but think it would be a lot easier just to take Prozac or fenfluramine. I've been reading so much about these drugs, I think of how easy it would be just to pop a pill and poof! I'd be thin in no time. Are they really so bad?

Answer: That depends on your point of view, your health, and your luck. Perhaps there are those women who benefit from using these drugs—those who are seriously overweight and are suffering from the dangerous effects morbid obesity can have on their physical and mental health. These drugs can be a life-saving miracle for these women.

Unfortunately, many other women used these drugs and now have serious lung and heart problems and will be nearly incapacitated for the rest of their lives. As of yet, there is no way to determine if you will be one of those women. Is it worth the risk? Also, these drugs are new, and many long-term side effects are still unknown. Do you really want to take a drug that alters your brain chemistry for an extended period of time? Also, there is evidence that many women regain the weight they've lost once they stop taking the drugs. Besides, food works just as powerfully as drugs and is safer. Why take drugs if you don't have to?

Question: What about oils? Can I use any oil I want?

Answer: Having a certain amount of fat is essential in your diet. (Unless, of course, your doctor has put you on a strict fat-free diet due to something like a heart condition.) But for most of us, especially women, 20 to 30 percent of the day's caloric intake should be predominantly unsaturated fat. Women who go on low-fat diets experience the symptoms of low serotonin. Be sure to put some of the right fat in your diet. Oils such as olive, canola, corn, sunflower, safflower, and sesame are an excellent source of unsaturated fats.

Question: You say this is a great way for women to eat. But I cook for my entire family. Is it okay for my husband to eat the same way I do? Are woman-friendly carbohydrates unfriendly to men?

Answer: Absolutely not. Woman-friendly carbohydrates like rice, barley, pasta, oatmeal, beans, and sweet potatoes are a great source of nutrition for both men and women. They just have a special advantage for women because they enhance the serotonin-making ability of women's brains. Many, many men have also lost weight easily with our program.

Question: Do you recommend any spaghetti sauce over another? Whenever possible I try to make my own, but for convenience sake I buy bottled brands.

Answer: Pretty much all of the bottled spaghetti sauces are fine to use except some of the low-fat or nonfat varieties. Check the label and be sure that sugar, corn syrup, fruit juice, fructose, or some other sweetener is not listed as one of the primary ingredients.

Question: I know that most breads are considered Bad Calories, but are there some combinations of ingredients that are okay to make bread with?

Answer: Sure. Breads that are made up of 75 to 80 percent whole grain *kernels* such as oats, barley, rye, wheat berries, and cracked wheat (bulgur) are considered Good Calories. It is important that the percentage of kernels in the recipe is this high. If you use a smaller percentage, say 50 percent kernels, you raise the bread's insulin-producing ability, making it a Bad Calorie. Guar gum also reduces the insulin-producing ability when used with carbohydrates. Our recipes for breads and muffins include it for this reason.

Question: Why is breakfast so important? Can't I just have a nonfat bagel—will it really be that bad?

Answer: Early morning is the time when it's easiest for a woman to use food to stimulate serotonin synthesis. Eating the right food for breakfast will produce significantly more serotonin than later in the day. Also, a woman's serotonin synthesis is naturally low in the morning. Eating Good Calorie carbohydrates for breakfast help her stimulate serotonin synthesis, making her function better through-out the day. Unfortunately, a bagel is not a Good Calorie because it releases too much insulin, thereby lowering serotonin. Eating a bagel for breakfast will contribute to food cravings later in the day.

Question: I need to lose thirty pounds. Once I achieve my goal how do I maintain it?

Answer: That will happen quite naturally if you continue to eat in a woman-friendly way and follow the six rules. This approach to food is healthy and sane, one that your body and mind will really enjoy. Most diets are something you suffer through to achieve your results. The effects of this program are quite different. You'll feel so good eating this way, you'll just continue because your body loves it.

Question: You have a chart of glycemic indexes in your book, but it doesn't jibe with some other charts I have come across. Why is that?

Answer: The figures we use are from the International Tables of Glycemic Indexes, which were published in the American Society for Clinical Nutrition in 1995. These indexes are updated periodi-cally, which accounts for some of the discrepancies. Also, we have included all the foods that have been tested; unfortunately not every food has been tested yet.

Question: I realize that it is better for a woman to eat less meat. When is the best time for me to eat my meal containing animal protein?

Answer: Lunch is the best time for a woman to eat meat. This gives her body enough time before the next morning's breakfast to elim-inate the protein from her body, as protein interferes with serotonin synthesis. This enables her body to create the serotonin she needs from the Good Calorie carbohydrates she's eating for breakfast.

However, this program is not fanatical. There will be times when this is impossible, but whenever it is possible you should eat your animal protein for lunch.

Question: I've been on the program for a few weeks but have not lost any weight. Is there anything I can do other than the caloric-restriction plan?

Answer: Check to be sure that you are following the rules, eating enough Good Calories for breakfast, and not snacking on Bad Calories. Review everything you're eating. Some women have more leeway to cheat with Bad Calories and are more sensitive to their effects. For example, Monika's serotonin chemistry is very sensitive, so her diet consists of predominantly Good Calories. That helps her stay slim, mentally astute, and emotionally strong. You might be like this too. Eat predominantly Good Calories for a week or two and keep Bad Calories to an absolute minimum, and see if you don't lose weight without having to go on the caloric-restriction program.

Question: Cooking rice and pasta—what should I know about that?

Answer: The key is *do not overcook*. Slightly undercooking both pasta and rice (this also applies to potatoes) significantly lowers their blood sugar–inducing ability. This makes them especially Good Calories. Besides, any Italian chef will tell you that *al dente* is the only right way to eat pasta!

Question: Can your diet help me overcome bulimia?

Answer: People with eating disorders have a hard time admitting to themselves and others that they have a problem. That you were able to admit your problem and ask for help is a great first step to healing. There seem to be a number of causes, both physical and psychological, associated with bulimia. Studies show, however, that bulimia is often accompanied by low serotonin and responds to treatment with Prozac and other serotonin-enhancing drugs. There is no harm in following the guidelines laid out in the free-eating program. However, this is not a substitute for the medical or psychological help you may require. Consult your physician before starting the program.

Philip consulted with one young woman who was five-foot-six and weighed ninety-six pounds. She phoned Philip in order to get confirmation of her mistaken belief that most foods were poisonous and should be avoided. Philip explained the principles of timing and food combination, and he convinced this young woman to try eating fully. He also convinced her to tell her doctor what she was doing and have him monitor her health. Over a period of a few months she gained weight until she weighed 116 pounds. She says that the only reason she could eat so many Good Calories is that they made her "feel so good." Perhaps this was a result of elevating serotonin. Without clinical studies, we make no claims. In her last conversation with Philip, this young woman said that her parents think that Good Calories saved her life.

Question: I'm a little confused. I thought caloric restriction was bad for a woman, yet you have a caloric-restriction plan.

Answer: Traditional caloric-restriction diets and low-fat diets create havoc in a woman's brain by depleting her of the essential nutrients to make serotonin. She needs sufficient serotonin for her brain to function normally. Without enough serotonin she opens herself up to a variety of symptoms: depression, food cravings, anxiety, irritability, inability to think clearly, mood swings, and more. However, when using our caloric-restriction program with predominantly Good Calories, the brain gets a sufficient amount of serotonin and prevents a woman's brain from being impaired so she doesn't experience those negative symptoms. She will also be able to stick with caloric restriction based on Good Calories because she will no longer have food cravings and will no longer be depressed.

Question: Do you have any information or comments on what to do or not to do when coming off a caloric-restriction diet? I was on Opti-Fast, which is 450 calories a day.

Answer: The best thing to do is immediately begin eating a diet of predominantly Good Calories. You have been on an extended period of semi-starvation, so your brain is undoubtedly suffering from serotonin impairment.

Question: I'm a little confused about what rice to use—parboiled, long grained, high amylose? Can you help?

Answer: Basically you want to avoid the types of rice that were cultivated especially to have a high sugar content. These are the low-amylose and glutinous rices. Other rices, such as basmati, were an intermediate stage into the development of high-sugar rices. They are not as bad, but still are not optimal. Most rice with a longer cooking time is fine. Most rices that are firm, not crunchy, after being cooked are fine.

Question: Is it best not to combine proteins with starches? Does this also mean it is not good to combine dairy with starches?

Answer: Breakfast is the time when it is most important not to combine dairy with starches. The reason for this is that dairy contains protein and you should avoid proteins during the breakfast meal. This means that the popular combination of breakfast cereal and milk should be avoided during breakfast. If you do find yourself having a bowl of cereal and milk, try to drink as little of the milk as possible.

Combining starches with proteins can result in increased fat formation. It also can bring together starches and the unsaturated animal fats associated with dairy and meats. This combination increases blood sugar and insulin, and therefore should be avoided.

This does not mean you can never have protein and starches in the same meal. However, a meal that consists only of protein and starches, such as meat and potatoes, should be avoided.

Question: I've just learned that I am pregnant. Will this program help prevent me from bingeing? I gained sixty pounds in two previous pregnancies.

Answer: If you are pregnant you must consult with your physician before changing your diet. Pregnancy is certainly not the time to undertake a weight-loss program.

It is not clear whether our diet will help prevent the cravings associated with pregnancy. There is sound theoretical reason for believing that it may help; however, we have only the experience of a few individuals who found that the program helped them during pregnancy and especially after pregnancy, when they were trying to regain their figure. If a woman is going to get the Starvation Response she certainly is susceptible during pregnancy and while nursing.

The best we can tell you is to try the program, but pay particular attention to making sure that you have enough vegetable proteins. Do not go into a state of protein depravation, but do not overeat animal protein. You can use vegetables to fulfill your need for protein by combining two portions of a grain, such as rice or pasta, with one portion of legumes, such as beans or peas.

Everyone should be aware that eating a diet high in Good Calories does not mean that you should ignore your protein needs. At the same time, it is not necessary to fill yourself with animal proteins. As in all things, moderation is the key.

Question: Is it okay to drink a little alcohol? Also, if you had to choose, which is better—beer, white wine, or red wine?

Answer: Alcohol is a Bad Calorie because it is a source of carbohydrate sugars. Because it produces a rush of blood sugar and insulin, it is also a potent stimulator of serotonin synthesis. The price you pay for this dramatic increase in serotonin is a simultaneous increase in insulin, appetite, and food cravings. Additionally, alcohol may produce only a short-term increase in serotonin that may be followed by a decrease in serotonin. This is why many people feel the need to consume another drink after a few hours have gone by.

The best way to drink alcohol is to hide it in a stomach full of Good Calorie carbohydrates. This means the best time to start a drink is after the meal has already started. Before-dinner cocktails create much more insulin than do drinks consumed later.

During the day following alcohol consumption it is important to eat lots of Good Calories to help with serotonin synthesis. It will help to alleviate the food cravings that occur.

Which form of alcohol is best? If I had to choose from among your three choices, I would choose red wine. Beer is high in maltose, which is a sugar with one of the highest known glycemic indexes. Additionally, white wine does not have the salicyliates, aspirinlike chemicals, and antioxidants that give red wine its ability to protect against heart attacks.

Question: What about Nutrasweet?

Answer: Most people can tolerate artificial sweeteners. However, 10 to 15 percent of the population respond to artificial sweeteners as if they were real sugar. The resulting insulin surge is the same as experienced with sugars. These people should not use artificial sweeteners while on our program. You can tell if you are such a person if drinking a soda sweetened with an artificial sweetener produces a decrease in energy such as would follow an insulin surge. If you get such a post-sugar low after eating an artificial sweetener there is a good chance that your body cannot tell the difference between real and fake sweets.

Question: I am a person who is underweight. Does that mean I should eat only "Bad Calories"?

Answer: Eating Bad Calories is a great way to put on weight if you want to put on fat, develop the insulin problems associated with obesity, and increase your food cravings for the wrong foods. Many body-building drinks are high in Bad Calories, forcing body builders to starve themselves in order to get rid of the fat created by these drinks.

The best way to put on more weight is to eat more frequently, but always eat a proper diet. Five or six small meals a day are better than two or three. Never eat more than your body can assimilate at any one sitting.

WRITE OR E-MAIL US

We are always interested in your experiences. Who knows—maybe your experiences will find their way into a future book. You can contact us at the following addresses.

We will try to post a list of the most frequently asked questions (FAQ) and their answers at our Web site. We cannot mail hard copies of that file as we anticipate the amount of correspondence will grow too large.

U.S. mail:
Experiences
20 Sunnyside Avenue
Suite 322
Mill Valley, CA 94941

E-mail
experiences@NatSlim.com

Web site
http://www.NatSlim.com

Good and Bad Calories by Category

Good Calories are in boldface. Bad Calories are in regular type. Although technically only carbohydrates can be Good Calories, we have included other items such as meat just to show you the best selection in each category. To be included in this list an item must be less than 30 percent fat. Bad Calories followed by an asterisk (*) have 31 to 40 percent fat and are suitable for occasional use.

BAKERY PRODUCTS

Wheat (white) flour is one of the greatest sources of Bad Calories, because it causes up to twice as much blood sugar as sugar does. There are almost no conventional bakery products that are acceptable on our program.

cake, angel food
cake, flan
cake, sponge
croissant
crumpet
Danish pastry
doughnut

muffins
muffins, corn
muffin, oatmeal mix (Quaker Oats)
pies, fruit and nut
pizza, cheese (Pillsbury)
waffles (Aunt Jemima,
    Quaker Oats)

BEVERAGES

As a general rule, juicing fruits or vegetables will only increase the blood sugar–producing potential of the ingredient. The physical forces involved in juicing a fruit or vegetable destroy the fibers that are essential for lowering the amount of blood sugar it produces. Therefore, if the fruit or vegetable is a Bad Calorie, it will be even worse as a juice.

alcoholic beverages, all
apple drink
apple juice, unsweetened
apple-cranberry drink
apple-cranberry juice, sweetened
**apple-cranberry juice,
    unsweetened**
apricot nectar
beer, light
beer, normal
berry drinks
carrot juice
cherry drink
cherry juice, unsweetened
cider, unsweetened
cocoa
fruit drinks
fruit punch
grape juice, red

grape juice, unsweetened
grapefruit juice, sweetened
grapefruit juice, unsweetened
lemonade
Lucozade
orange cordial
orange drink
orange juice, sweetened
orange juice, unsweetened
papaya nectar
passion fruit juice
peach nectar
pineapple juice, sweetened
pineapple juice, unsweetened
raspberry-cranberry drink
soft drinks, diet
soft drinks, regular
tomato juice

BREADS

biscuit
barley flour bread
barley kernel (80 percent kernel)
barley kernel (50 percent kernel)
bread stuffing
**bulgur bread (75 percent
    cracked wheat kernels)**
bulgur bread (50 percent
    cracked wheat kernels)

croutons
croutons, cheese
French baguette
hamburger bun
kaiser roll
matzo
matzo meal
**oat bran bread
    (50 percent oat bran)**

BREADS *(continued)*

oat bran bread
  (45 percent oat bran)
pita bread, white
rolls
rye bread, cocktail sliced
rye flour bread
rye flour bread, whole meal
rye kernel bread
  (80 percent kernels)
rye kernel bread,
  hard pumpernickel

rye kernel pumpernickel,
  whole grain
wheat, puffed crisp bread
wheat bread
wheat bread, gluten-free
wheat bread, semolina
wheat bread, whole-meal flour
white flour bread
white flour, high-fiber
Whole Wheat Snack Bread
  (Ryvita)

BREAKFAST CEREALS

Most breakfast cereals are Bad Calories.

Oat cereals are one of the few types of Good Calorie cereals. However, it is important that they not be overly processed. For example, instant oatmeal is unacceptable, whereas regular oatmeal is a Good Calorie.

All-Bran (Kellogg's)
Bran Buds (Kellogg's)
Bran Buds with psyllium
  (Kellogg's)
Bran Chex (Kellogg's)
Cheerios (General Mills)
Cocopops (Kellogg's)
Corn Bran
Corn Chex
cornflakes
Corn Flakes (Kellogg's)
Cream of Wheat (Nabisco)
Cream of Wheat, instant (Nabisco)
Crispix (Kellogg's)
Golden Grahams (General Mills)
granola
Grape-Nuts Flakes (Post)
Grape-Nuts (Post)
hominy

Life (Quaker Oats)
Müeslix
oat bran, flaked
oat bran, raw
oat bran cereal
oat bran flakes
oatmeal, instant
oatmeal, long cooking
oatmeal, quick
Oats, One Minute (Quaker Oats)
oats, porridge
Oats, Quaker Quick
oats, rolled
Pro Stars (General Mills)
puffed rice
puffed wheat
Puffed Wheat (Quaker Oats)
rice bran, raw
Rice Bubbles (Kellogg's)

Rice Chex (Nabisco)
Rice Krispies (Kellogg's)
shredded wheat
shredded wheat, Mini Wheats
   (Kellogg's)
Shredded Wheat (Nabisco)
Special K (Kellogg's)

Sultana Bran (Kellogg's)
Sustain (Kellogg's)
Team (Nabisco)
Total (General Mills)
wheat, flaked
Vita-Bits
Wheetabix

CEREAL GRAINS

barley
barley, pearled
barley, cracked
barley, rolled
buckwheat
bulgur
cornmeal
corn taco shells (Old El Paso)
couscous
millet
rice, brown
rice, white, Cajun Style
   (Uncle Ben's)
rice, white, basmati
rice, white, Garden Style
   (Uncle Ben's)
rice, white, high amylose
rice, white, instant

rice, white, long grain
rice, white, Long Grain & Wild
   (Uncle Ben's)
rice, white, low amylose
rice, white, Mexican Fast and
   Fancy (Uncle Ben's)
rice, white, parboiled
rice, white, parboiled,
   Uncle Ben's converted
rice, white, parboiled,
   Uncle Ben's long grain
rice, white, short grain
rice, white, waxy
rye, whole kernel
tapioca
wheat, whole kernels
   (wheat berries)
wheat kernels, quick cooking

COOKIES

digestive biscuits
Graham Wafers (Nabisco)
oatmeal

rich tea
shortbread
vanilla wafers

CRACKERS

crackers, plain
crisp breads
melba toast
Premium Soda Crackers (Nabisco)
rice cakes
rye crispbread
rye crispbread, high-fiber

rye crispbread, high-fiber
   (Ryvita Co.)
rye crispbread, high-fiber
   (Kavli Norwegian Crispbread)
Stoned Wheat Thins (Nabisco)
water crackers

CONDIMENTS

The only key to these products is their fat content. Very few have high blood sugar–producing potential.

Mayonnaise should be avoided or used sparingly. Even diet mayonnaise is still too high in fat content, and the nondairy mayonnaise made with tofu or soybean oils are Bad Calories.

barbecue sauce
béarnaise sauce
carob chips
carob powder

mayonnaise, regular
mayonnaise, light
mustard
salsa

DAIRY PRODUCTS

In this category, fat content is the problem. However, just because a dairy product has a high fat content does not mean you cannot use small amounts for cooking. Use high-fat dairy products sparingly.

butter
butterfat
cheese, high fat
cheese, low fat
cottage cheese, high fat
cottage cheese, low fat
cream, all
cream cheese, all
custard, milk

eggs, chicken
egg whites, chicken
eggs, duck
eggs, quail
ice cream, high fat
ice cream, some low fat
margarine
milk, high fat
milk, 1 or 2 percent fat

milk, skim

milk, 1 or 2 percent fat, chocolate, sugar sweetened

milk, 1 or 2 percent fat, chocolate, artificially sweetened

sherbet

sorbet

tofu frozen dessert, nondairy

Vitari nondairy frozen fruit product

yogurt, low fat or nonfat

yogurt, low fat or nonfat, artificially sweetened

yogurt, high fat or sweetened

FISH

The choices marked with an asterisk (*) indicate a marginally high fat content and should be used sparingly.

abalone*

alewife (herring)*

anchovy

anchovy, canned

bass, black

bass, freshwater

bass, striped

bluefish*

butterfish

carp

carp, roe

carp, smoked*

catfish

chub

clams, canned

clams, steamed

cod

crab, deviled

crab, king

crab, soft shell

crayfish

cuttlefish

dock

dogfish

eel

flounder

gefilte fish, in broth

grouper

haddock

halibut

halibut, smoked

herring

herring, kippered

lobster*

mackerel

monkfish

mussel

octopus

oyster*

perch

pike

pompano

rockfish

sable fish

salmon, Atlantic

salmon, fish farm

salmon, pink, canned*

sardines

scallops

scrod

sea bass

sea trout*

shad

shark*

FISH *(continued)*

sheepshead	**snapper**
shrimp	**sole**
shrimp, canned	tuna, canned in oil
smelt	**tuna, canned in water**

FRUIT, FRESH

Avoid overly ripe fruits. The riper the fruit, the more blood sugar–producing potential it has.

apple	honeydew melon
apricot, dried	**kiwifruit**
apricot, fresh	**lemon**
banana, ripe	**lime**
banana, underripe	**mandarin orange**
blackberry	mango
blueberry	**nectarine**
boysenberry	**orange**
cantaloupe	papaya
casaba melon	**peach**
cherry	**pear**
cranberry	pineapple
date	**plum**
fig, dried	prune
fig, fresh	**strawberry**
grape	raisin
grapefruit	watermelon

FRUIT, CANNED OR IN JARS

applesauce, sweetened	cherries, in syrup
applesauce, unsweetened	**cherries, in water**
apricot, natural	**cherries, maraschino,**
blackberries, in syrup	**unsweetened**
blackberries, in water	**cranberry sauce, unsweetened**
blueberries, in water	figs, in syrup
blueberries, sweetened	**figs, in water**

fruit cocktail, canned
  (Delmonte)
fruit cocktail, in syrup
fruit cocktail, in water
fruit salad, in syrup
fruit salad, in water
jam and jelly, sweetened
jam and jelly, unsweetened
mandarin orange, in syrup

mandarin orange, in water
peaches, canned, light syrup
peaches, sweetened
peaches, unsweetened
pears, sweetened
pears, unsweetened
plums, sweetened
plums, unsweetened
prunes

FRUIT, FROZEN

Avoid any frozen fruit packaged in syrup.

blackberries, in syrup
blackberries, in water
blueberries, in water
blueberries, sweetened
cherries, in water

raspberries, in water
rhubarb, in water
rhubarb, sweetened
strawberries, sweetened
strawberries, in water

LEGUMES

Legumes are one of the best Good Calories. They often have less than half the blood sugar–producing potential of sugar. By adding them to a meal, you can drastically reduce that meal's fat-forming potential.

Combining half a unit of legumes to one unit of grain (e.g., ½ cup beans and 1 cup of rice) produces a perfect protein. If you are trying to maximize your weight loss by avoiding meat products, this combination is a great substitute.

adzuki beans, dried
adzuki beans, canned
baked beans, canned (Libby's)
bean dip
black bean soup
black-eyed peas, dried
butter beans
chickpeas, canned
chickpeas, dried
chili beans

dhal lentils
green peas, dried
green peas, frozen
haricot beans
kidney beans, canned
kidney beans, dried
lentils, green, canned
lentils, green, dried
lentils, red, dried
lima beans, baby, frozen

LEGUMES (continued)

lima beans, dried   **pinto beans, dried**
moth beans   refried beans, canned
mung bean sprouts   **refried beans, canned, low fat**
navy beans, white, dried   **romano beans**
peanuts   **soybeans, dried**
peanut butter   **split peas, yellow, dried**
pinto beans, canned   **white beans, haricot, dried**

MEAT

BEEF

ribs   rib roast
blade   ribs
brisket, lean   round, bottom, no fat*
chuck, no fat   round, eye, no fat*
club steak lean*   round, tip, no fat*
corned beef   round, top, no fat
flank, lean   rump roast, no fat*
ground, lean   sirloin, lean*
kidneys   T-bone
liver*   tenderloin
porterhouse, no fat   tongue
ribeye   top loin

BISON

all forms

LAMB

chops   loin
kidneys   shank
leg, roast   shoulder
liver   sirloin

LIVER

beef   **pork**
lamb   **veal**

PORK

arm, lean roasted
ham, extra lean
heart
kidney
liver

loin
ribs
tenderloin, lean
tenderloin, roasted, lean

RABBIT

wild or domestic

VEAL

breast
chuck
chop
chop, lean only
cutlet, lean only
heart
kidneys

liver
loin
ribs
round
rump
sweetbreads
tongue

NUTS

acorns
acorn flour
almond butter
almond powder, defatted*
almond powder, normal
almonds
Brazil nuts
cashews
cashew butter
chestnuts
coconut milk
coconut water

coconut, sweetened
coconut, unsweetened
coconut, whole
filberts
macadamias
peanuts
peanuts, dry roasted
pecans
pine nuts
pistachios
walnuts

PASTA AND PASTA SAUCES

Pasta remains a Good Calorie as long as it is not overcooked. The pasta should be firm, not mushy and starchy.

capellini
clam sauce, red
clam sauce, white
fettuccine, egg-enriched
linguine
macaroni
Macaroni & Cheese, boxed
    (Kraft)
noodles, egg

rice pasta, brown, low-amylose
    rice flour
spaghetti, durum
spaghetti, protein enriched
spaghetti, white
spaghetti, whole meal
spaghetti, whole wheat
star pasta
vermicelli

POULTRY

CAPON

dark meat, with skin
white meat, with skin

whole

CHICKEN

back, no skin
breast with skin*
breast, no skin
breast, smoked*
chicken roll, light
chicken salad
dark meat, no skin
drumstick, no skin*
fried

leg, no skin
luncheon meat
luncheon meat, white*
roasting, whole*
stewing, whole
thigh, no skin
white meat, no skin
wing, no skin*
wing, with skin

DUCK

domesticated
liver*

wild, meat only*

GAME

guinea hen, meat only (no skin)
guinea hen, whole*

pheasant
quail, breast
quail, whole

GOOSE

all forms

TURKEY

breast, fillet, no skin
breast roast*
breast slices
breast, slices, no skin
breast tenderloin
back meat only*
dark meat, no skin
dark meat, with skin
drumsticks, with skin
drumsticks, no skin*
giblets, simmered

gizzard, simmered
heart, simmered
light meat, no skin
light meat, with skin*
liver, simmered
thighs, no skin
thighs, with skin
wing drumettes, no skin*
wings, no skin
wings, with skin

SOUPS

black bean (Wil Pak)
green pea, canned (Campbell's)
lentil, canned

split pea (Wil Pak)
tomato

SUGARS

fructose (see warning against use on p. 85)
glucose
honey

lactose
maltose
molasses
sucrose (table sugar)

VEGETABLES, FRESH

acorn squash
alfalfa sprouts
aloe vera juice
artichoke
asparagus
aubergine
avocado
avocado dip
bamboo shoots
bean sprouts
beet greens
beetroot
beets
broccoli
brussel sprouts
butternut squash
cabbage
carrot
cauliflower
celery
coleslaw
corn, sweet
cucumber
eggplant
endive
green beans
Jerusalem artichoke
kale
leek
lettuce

mushroom
mustard greens
okra
olive
onion
parsnip
peas, fresh
peas, frozen
peppers, bell, green
peppers, bell, red
potato, baked without fat
potato, french fried
potato, microwaved
potato, new, undercooked
potato, new, well cooked
pumpkin
radish
rhubarb
rocket
rutabaga
spinach
spring greens
squash, green
squash, yellow
string beans
sweet potato
tomato
turnip
yam
zucchini

VEGETABLES, CANNED

artichoke hearts, in water
artichoke hearts, marinated,
    in oil
asparagus
baby food
bamboo shoots
bean salad
beans
beets
butter bean
butter beans, with pork*
Chinese vegetables
collard greens
corn

cucumbers, pickled,
    unsweetened
green beans
kale
mushrooms
mustard greens
okra
olives
onion
peas
pickles
potato
sauerkraut
tomato paste, unsweetened

Glycemic Index of Selected Foods

(Good Calories in boldface type)

BAKERY PRODUCTS

Cake

angel food	95
banana, made without sugar	79
flan	93
sponge (high in protein, not for breakfast)	**66**

croissant 96
crumpet 98
doughnut 108

muffins

corn, low amylose (sweet corn)	146
Oatmeal muffin mix (Quaker Oats)	98

pizza, cheese (Pillsbury) 86
waffles (Aunt Jemima, Quaker Oats) 109

BEVERAGES

cordial, orange 94
Lucozade 136
soft drink, Fanta (Coca-Cola Bottlers) 97

BREADS

bagel, white	103
barley flour bread	95
barley kernel bread	
80 percent kernels	**54**
50 percent kernels	**66**
bread stuffing, Paxo (Campbell Soup Co.)	106
bulgur bread (cracked wheat)	
75 percent cracked wheat kernels	**69**
50 percent cracked wheat kernels	83
French baguette	136
hamburger bun	87
kaiser roll	104
melba toast	100
oat bran bread	
50 percent oat bran	**63**
45 percent oat bran	**72**
oat kernel bread, 80 percent kernels	93
rye flour bread	92
whole meal	89
rye kernel bread	
80 percent kernels	**66**
pumpernickel	**58**
whole-grain pumpernickel	**66**
cocktail, sliced	84
wheat bread	
white flour	101
white flour, high-fiber	97
wheat bread, gluten free	
whole meal flour	100
whole wheat snack bread (Ryvita)	105
pita bread, white	82
semolina bread	92

BREAKFAST CEREALS

All-Bran (Kellogg's)	**43**
Bran Buds (Kellogg's)	83
Bran Buds with psyllium (Kellogg's)	**67**
Bran Chex (Kellogg's)	83
Cheerios (General Mills)	106

BREAKFAST CEREALS *(continued)*

Cocopops (Kellogg's)	110
Corn Bran	107
Corn Chex	118
Corn Flakes (Kellogg's)	110
corn flakes	122
Cream of Wheat (Nabisco)	94
Cream of Wheat, instant (Nabisco)	105
Crispix (Kellogg's)	124
Golden Grahams (General Mills)	102
Grape-Nuts (Post)	96
Grape-Nuts Flakes (Post)	114
Life (Quaker Oats)	94
Müeslix	94
Nutri-grain (Kellogg's)	94
oat bran, raw	**78**
oatmeal, quick	
Quaker Quick Oats (Quaker)	93
One Minute Oats (Quaker)	94
oats, long cooking	**49**
oats, porridge	89
oats, rolled	85
Pro Stars (General Mills)	102
puffed rice	132
puffed wheat (Quaker Oats)	96
rice bran, raw	**27**
Rice Bubbles (Kellogg's)	126
Rice Chex (Nabisco)	127
Rice Krispies (Kellogg's)	117
shredded wheat	
Mini Wheats (Kellogg's)	83
Shredded Wheat (Nabisco)	118
Special K (Kellogg's)	**77**
Sultana Bran (Kellogg's)	**74**
Sustain (Kellogg's)	97
Team (Nabisco)	117
Total (General Mills)	109
wheat, flaked	
Vita-Bits	87
Wheetabix	106

CEREAL GRAINS

barley	
barley	**35**
barley, pearled	**37**
cracked	**72**
rolled	94
buckwheat	**78**
bulgur, boiled 20 minutes	**68**
cornmeal	98
taco shells (Old El Paso)	97
couscous	93
millet	101
rice, brown	**79**
rice, white	
basmati	83
instant	128
high amylose	91
low amylose	83
long grain	**73**
parboiled rice	**68**
Uncle Ben's converted, boiled 20 to 30 minutes	**54**
Uncle Ben's converted, long grain, boiled 20 to 30 minutes	**64**
waxy	126
specialty rices	
Cajun Style (Uncle Ben's)	**72**
Garden Style (Uncle Ben's)	**79**
Long Grain & Wild (Uncle Ben's)	**77**
Mexican Fast and Fancy (Uncle Ben's)	83
rye, whole kernel	**48**
tapioca, boiled with milk	115
wheat, whole kernel	**59**
wheat kernels, quick cooking	**75**

COOKIES

digestive biscuits	82
Graham Wafers (Nabisco)	106
oatmeal	**77**
rich tea	**79**

COOKIES *(continued)*

shortbread   91
vanilla wafers   110

CRACKERS

rice cakes   117
rye crispbread, high-fiber
   Ryvita (Ryvita Co.)   84
   rye crispbread   90
   Kavli Norwegian Crispbread   101
Stoned Wheat Thins (Nabisco)   96
water crackers   90
   Premium Soda Crackers (Nabisco)   106

FRUIT AND FRUIT PRODUCTS

apple   **53**
   Golden Delicious   **56**
   Braeburn   **45**
apple juice, unsweetened   **57**
applesauce, unsweetened   **41**
apricots, canned, light syrup   91
apricots, dried   **44**
banana   91
banana, underripe   **43**
cherries   **32**
fruit cocktail, canned (Delmonte)   **79**
grapefruit   **36**
grapefruit juice, unsweetened   **69**
grapes   **62**
kiwifruit   **75**
mango   80
orange   **62**
orange juice   **74**
papaya   83
peach, canned, natural juice   **43**
peach, canned, heavy syrup   83
peach, canned, light syrup   **74**
peach, fresh   **40**
pear   **51**
pineapple   94

plum	**34**
raisins	91
rock melon	93
sultana	80
watermelon	103

LEGUMES

baked beans, canned (Libby's)	80
black-eyed peas, dried	**59**
butter beans	**44**
chickpeas, canned	**60**
chickpeas, dried	**47**
kidney beans, canned	**74**
kidney beans, dried	**49**
lentils, green, canned	**74**
lentils, green, dried	**42**
lentils, red, dried	**36**
lima beans, baby, frozen	**46**
lima beans, dried	**36**
navy beans, white, dried	**54**
pinto beans, canned	**64**
pinto beans, dried	**55**
romano beans	**65**
soybeans, canned	**22**
soybeans, dried	**20**
split peas, yellow, dried	**45**

PASTA

capellini	**64**
fettuccine, egg-enriched	**46**
linguine	**68**
macaroni	**64**
Macaroni & Cheese, boxed (Kraft)	92
rice pasta, brown, low-amylose rice flour	131
spaghetti, durum	**78**
spaghetti, protein enriched	**38**
spaghetti, white	**56**
spaghetti, whole meal	**61**
star pasta	**54**
vermicelli	**50**

ROOT VEGETABLES

beets	91
carrots	101
parsnip	139
potato, baked without fat	121
potato, french fried	107
potato, instant	118
potato, microwaved	117
potato, new, well cooked	100
potato, new, undercooked	**67**
rutabaga	103
sweet potato	**77**
yam	**73**

VEGETABLES

peas, dried	**56**
peas, green	**77**
peas, green, frozen	**68**
pumpkin	107
sweet corn	82

All other vegetables are Good Calories.

SUGARS

honey	104
fructose (see warning against use)	**32**
glucose	138
lactose	**65**
maltose	158
sucrose	84

SOUPS

black bean (Wil Pak)	92
green pea, canned (Campbell's)	94
lentil, canned	**63**
split pea (Wil Pak)	86
tomato	**54**

ETHNIC FOODS AND MEALS

Asian Indian Foods

baisen chapati	**39**
bajra chapati	**82**
barley chapati	61
bengal gram dal, chickpeas	16
black gram	**61**
black gram dal	**66**¦
green gram	**54**,
green gram, whole	81
green gram dal	89
horse gram	**73**
jowar	110
Lentils & Rice (East Indian Foods, frozen)	81
maize, chapati	89
ragi	123
rajmah	**27**
semolina, preroasted	109
semolina, steamed	**79**
tapioca, steamed	100

Australian Aboriginal Foods

mulga seed	**11**
bunya nut pine	**67**
bush honey, sugar bag	**61**
blackbean seed	**11**
cheeky yam	**49**

Chinese Foods

Chinese glutinous rice	98
rice vermicelli	83
lungkow bean thread	**37**

Mexican Foods

black beans	**43**
brown beans	**54**
nopal, prickly pear cactus	**10**

Pacific Islands Foods

breadfruit	97
taro	**77**
sweet potato	**63**

ETHNIC FOODS AND MEALS *(continued)*

Pima Indian Foods

acorns, stewed with venison	**23**
cactus jam	130
corn hominy	**57**
fruit leather	100
lima-bean broth	**51**
mesquite cakes	**36**
white teparies broth	**44**
yellow teparies broth	**41**

South African Foods

brown beans	**34**
gram dal	**7**
maize meal porridge, unrefined	101
maize meal porridge, refined	106
m'fino, wild greens	97

DAIRY FOODS

ice cream	51–114
milk	
full-fat	**39**
skim	**46**
chocolate, sugar sweetened	**49**
chocolate, artificially sweetened	**34**
custard, milk	**61**
tofu frozen dessert, nondairy	164
vitari, nondairy, frozen fruit product	**40**
yogurt	
low-fat, fruit, sugar sweetened	**47**
low-fat, artificial sweetener	**20**

Mail-Order Resources

BREAKFAST FOODS, BREADS, DRINKS, AND BARS

As we write these words, we are talking to several firms that tell us they will be manufacturing these Good Calorie foods. At this time we cannot tell if they will carry forth with their promises. We are aware only of a low-protein Good Calorie drink, suitable for mornings. So, as a service to our readers, we have decided to establish these ways you can contact us and obtain information on how to order drinks and snacks.

The site will also contain the latest medical discoveries, book updates, questions and answers, sample chapters, recipes, discussion forums, and scientific abstracts. You will be able to post questions which we will answer.

U.S. mail:
Good Calorie Products
20 Sunnyside Avenue
Suite 322
Mill Valley, CA 94941

Telephone: NAT-SLIM (888-628-7546) (toll free)

E-mail: info@NatSlim.com

Internet: http//www.NatSlim.com

YOGA

There are Iyengar Yoga classes in many cities throughout the United States. Look in the yellow pages under "Yoga" for classes in your area. There is a difference between yogas, and we recommend Iyengar Yoga.

If you live in the San Francisco Bay area we recommend attending one of Manouso Manos's classes. He is an extraordinary yoga instructor—it's an experience not to be missed. Beginners are most welcome. He is one of the most capable and experienced Senior Iyengar teachers in the United States. He also travels throughout the United States giving weekend workshops. For his schedule, write the address listed below.

We know firsthand about the magic of Manouso's touch. He has been the sole reason that Philip is recovering from a serious automobile accident. Philip turned to Manouso when all else failed, after the specialists said his neck was inoperable because so many disks had been damaged and that he had better learn to live on painkillers and go through life in an iron neck brace. Manouso Manos and the Iyengar Yoga system has given Philip his health back. Naturally, we recommend them highly.

Manouso Manos
236 West Portal Avenue, Suite 196
San Francisco, CA 94127

Manouso Manos's *Yoga: The Home Study Course:*
Vol. I The Standing Poses
Vol. II The Inversions
Vol. III Forward Bends

Any one tape, $39.95 and $3 shipping
Any two tapes, $75 and $4 shipping
All three tapes, $100 and $5 shipping
(California residents add applicable sales tax.)

APPENDIX IV

Recipes

SNACKS AND APPETIZERS

Sweet Potato Fries

2 large unpeeled sweet potatoes,
    cut into thin wedges
vegetable oil spray

½ teaspoon salt
1 teaspoon Cajun spice
    (optional)

Preheat oven to 350°. In a medium bowl, spray potato wedges with oil spray; toss with salt and Cajun spice, if desired. Place wedges in a single layer on a baking sheet.

Bake until tender, about 20 to 25 minutes.

Makes 4 servings.
115 calories per serving.

Dry-roasted Chickpeas

There is an interesting story behind this recipe. A woman from Scarsdale, New York, wrote: "For the past thirty-five years, I have agonized over keeping my dress under size 12. After only four months of starting your program, I now wear a size 8 and am eating the way I always wanted to and never thought I could." She offered this great piece of advice:

"One of my tricks for anticipating a Bad Calorie is to always have a soaked or roasted chickpea around, to pop in my mouth with a drink or another Bad Calorie. I carry them in my pocket if I go to a cocktail party where I might be tempted to pick up a nut or a cracker."

What a great idea! All legumes are good calories. Whenever we add them to other foods they make Bad Calories less bad. We offer two delicious recipes for roasted chickpeas.

2 cups cooked chickpeas   ½ teaspoon salt
  (garbanzo beans)
½ teaspoon garlic powder or 1
  to 2 garlic cloves, crushed

Preheat oven to 350°. Mix all ingredients together and spread on a greased pan or cookie sheet, making sure there is only a single layer of beans on the pan. Cook for 30 minutes or until chickpeas are crisp on the outside, tender on the inside.

Makes 2 cups.
134 calories per ½ cup.

Dry-roasted Curried Chickpeas

This is a great snack or topping for salads.

2 cups cooked chickpeas   ½ to 1 teaspoon salt
  (garbanzo beans)   ½ teaspoon turmeric
1½ teaspoons coriander   ½ teaspoon garlic powder
1 teaspoon curry powder   (optional)

Preheat oven to 350°. Mix all ingredients together and spread on a greased pan or cookie sheet, making sure there is only a single layer of beans on the pan. Cook for 30 minutes or until chickpeas are crisp on the outside, tender on the inside.

Store in a sealed container or plastic bag. You can crumble these onto salads for a delicious, low-fat topping.

Makes 2 cups.
134 calories per ½ cup.

Spicy Bean Dip

1 cup kidney beans, cooked
1 cup tomato, chopped fine
1 tablespoon olive oil
½ to 1 teaspoon chili powder

½ to 1 teaspoon cumin
dash cayenne pepper
salt to taste

Puree beans in a food processor or blender, or mash with fork. Add the rest of the ingredients; puree or mash until well blended.

Makes about 1 cup.
98 calories per ¼ cup.

Hummus

2 cups cooked chickpeas
 (garbanzo beans)
2 large garlic cloves (optional)
½ cup plain nonfat yogurt
juice of ½ lemon (to taste)

3 tablespoons tahini*
2 tablespoons olive oil
½ teaspoon cumin
salt to taste

Puree chickpeas and garlic in a food processor or blender. Add yogurt, lemon juice, tahini, olive oil, cumin, and salt. Blend thoroughly until smooth. Adjust seasonings to taste.

Serve with raw vegetables.

Makes 2½ cups.
100 calories per ¼ cup.

*Tahini is ground sesame seeds made into a butter, like peanut butter. It is available in health food and specialty stores.

Eggplant Dip

3 medium eggplants
2 teaspoons olive oil
6 tablespoons plain nonfat
 yogurt
4 tablespoons lemon juice

3 tablespoons parsley
1 to 2 garlic cloves, minced
 (optional)
¼ to ½ teaspoon cumin (optional)
salt and pepper to taste

Preheat broiler. Place eggplants on oiled baking sheet, about 4 inches under the broiler. Broil until eggplants are charred and soft, turning every 5 minutes or so. Remove from heat and allow to cool.

Remove charred skin from eggplants and squeeze out liquid from flesh very carefully. Discard liquid and charred skin. Blend eggplants and all other ingredients until very smooth. Cover and refrigerate for a few hours. Serve with raw vegetables.

Makes about 2 cups.
88 calories per ¼ cup.

Curried Yogurt Dip

¾ cup plain nonfat yogurt
2 teaspoons curry powder
1 teaspoon lemon juice

¼ teaspoon sugar
¼ teaspoon black pepper
few drops hot pepper sauce

In a bowl, combine ingredients, stirring well. Adjust seasonings to taste. Serve with raw vegetables.

Makes ¾ cup.
22 calories per ¼ cup.

Yogurt Cheese

A delicious substitute for cream cheese,
crème fraîche, or heavy cream.

1 quart plain nonfat yogurt
herbs (optional)

garlic powder (optional)
onion powder (optional)

Line a fine strainer or colander with cheesecloth or a coffee filter and set it on top of a bowl for drainage.

Pour yogurt into the strainer and refrigerate for at least 4 hours or as long as overnight. Discard liquid in bowl.

Add any combination of herbs and garlic, if desired.

Note: The longer you drain the yogurt the firmer the curds.

Makes 2 cups.
110 calories per cup.

Elegant Fish Pâté

This fish pâté and the vegetable pâté are wonderful to serve for an appetizer or part of a main course for an elegant lunch or dinner.

⅓ pound diced salmon
⅓ pound shrimp
⅓ pound crab meat (or lobster)
8 ounces scallops
8 ounces flounder or sole fillets
½ cup heavy cream
3 egg whites, beaten

1 tablespoon basil*
1 tablespoon chives*
1 tablespoon parsley*
1 tablespoon tarragon*
¼ teaspoon nutmeg
black or white pepper to taste
vegetable seasoning to taste

For best results chill the food processor blade, bowl, and mixing bowl before beginning as well as the fish, cream, and eggs.

Preheat oven to 225°. Lightly oil a 6-inch loaf pan.

Put all of the fish in food processor and puree for 45 seconds. Add cream *very* slowly. Transfer to clean, chilled bowl. Stir in the rest of the ingredients. Blend well.

Pour into loaf pan; wrap tightly in foil. Place loaf pan into a larger, shallower pan. Fill shallower pan with very hot water so that it comes halfway up the pâté pan.

Bake pâté for 1 hour.

Let cool completely at room temperature. Then chill until cool, about 2 hours.

Unmold pâté and cut into 4 slices.

Serve over lettuce with a slice of lemon.

Makes 4 servings.
311 calories per serving.

*Use fresh herbs if possible—if using dried herbs, reduce quantity by half.

Vegetable Terrine

½ pound fresh spinach
½ pound Swiss chard
1 red onion, finely chopped
2 tablespoons olive oil
nutmeg to taste

black or white pepper to taste
vegetable seasoning to taste
4 egg whites, beaten
¼ cup low-fat ricotta cheese
¼ cup Parmesan cheese

Preheat oven to 350°. Lightly grease a 6-inch loaf pan.

Remove stems from spinach and chard. Wash thoroughly, pat dry, and chop coarsely.

Sauté onion in olive oil until soft, about 5 minutes. Add chard and spinach and toss for 4 minutes. Add nutmeg, pepper, and vegetable seasoning. Turn into large mixing bowl and allow to cool.

Add egg whites, ricotta, and Parmesan to spinach-and-chard mixture. Blend well. Pour into the loaf pan and bake for 45 minutes. Allow to cool for 10 minutes before unmolding loaf. Decorate with thinly sliced raw vegetables.

Serve as an appetizer or with rice as a main course.

Makes 4 servings.
220 calories per serving.

SOUPS

Quick Garlic Soup

This soup can be used as a soup stock.

5 cups water
4 to 6 garlic cloves, minced or
   put through a press
½ teaspoon ground sage
½ teaspoon dried thyme

salt to taste
1 bay leaf
2 tablespoons fresh parsley,
   chopped

Bring water to a boil. Add garlic, sage, thyme, salt, bay leaf, and parsley. Simmer 15 minutes. Adjust seasonings. This is a basic stock, and can be reheated.

Note: For an additional variation, see Egg-Lemon Soup, which follows.

Makes 5 cups.
6 calories per serving.

Egg-Lemon Soup

1 quart Quick Garlic Soup
   (recipe above)
4 egg whites
½ cup lemon juice

salt and pepper to taste
1 cup cooked brown or white rice
parsley
lemon slices

Bring Quick Garlic Soup to a simmer.

Beat egg whites in a bowl, then beat in lemon juice. Stir some of the hot soup into egg mixture. Transfer mixture to soup pot. Heat, but do not boil, to prevent eggs from curdling. Add salt and pepper to taste.

Divide rice among 4 bowls. Ladle soup on top of rice. Sprinkle with parsley and garnish with a lemon slice.

Makes 4 servings.
68 calories per serving.

Gazpacho

4 ripe tomatoes, peeled and diced
1 small onion, very finely diced
1 cucumber, peeled and very
   finely diced
1 green bell pepper, very finely diced
1 red bell pepper, very finely diced
½ carrot, very finely diced
2 garlic cloves, minced (optional)

10 fresh basil leaves, slivered
2 tablespoons parsley, finely
   chopped
1 tablespoon olive oil
¼ to ½ cup fresh lemon juice
2 cups cold tomato juice
salt and pepper to taste
red pepper sauce (optional)

Mix all ingredients together.

Makes 4 servings.
115 calories per serving.

Quick Egg Drop Soup

1 scallion, sliced
2 cups chicken broth

2 egg whites

In medium saucepan, add scallion to chicken broth and simmer gently. Beat egg whites and add to chicken broth very slowly, while stirring soup rapidly so that egg develops into noodlelike strands. Cook only until eggs are firm.

Makes 2 servings.
70 calories per serving.

Quick Vegetable Soup

1½ cups water
½ cup parsnips or turnips, chopped
¼ cup green beans, chopped
½ cup Chinese cabbage, shredded
½ cup zucchini, chopped
½ cup yellow squash, chopped
½ cups fresh parsley, chopped

1 medium tomato, chopped
½ teaspoon dried thyme
½ teaspoon dried rosemary
½ teaspoon dried marjoram
vegetable seasoning
salt to taste

In a medium saucepan, bring water to a boil. Add the parsnips and green beans; simmer 4 to 5 minutes.

Add the cabbage, zucchini, and squash, and simmer for a few minutes more.

Add the parsley, tomato, thyme, rosemary, marjoram, vegetable seasoning, and salt. Simmer a few more minutes.

Remove from heat and adjust seasoning.

Makes 2 servings.
65 calories per serving.

Rice Noodle Soup

4 stalks lemongrass
6 cups chicken broth
3 scallions, sliced diagonally
½ cup fresh cilantro, chopped

⅓ cup tamari or soy sauce
1 to 1½ teaspoons fresh ginger, finely chopped
6 ounces of rice noodles

Cut off the stalks of the lemongrass, then slice the rest thinly. (Use the tops for salads or other vegetable dishes.)

In a large saucepan, combine the lemongrass, chicken broth, and scallions. Bring to a gentle boil. Add the cilantro, tamari, and ginger; simmer, uncovered, for about 30 minutes.

Add the noodles, stirring, and cook for another 2 minutes, being careful not to overcook the noodles.

Makes 6 servings.
138 calories per serving.

Vegetable and Pasta Soup

2 tablespoons olive oil
1 medium onion, finely chopped
8 ounces mushrooms
1 teaspoon each basil, oregano, and marjoram
6 cups chicken or vegetable broth or Quick Garlic Soup (see page 185)

1 banana squash, peeled and cut into ½-inch cubes
1 medium thin-skinned potato, peeled and cut into ½-inch cubes
¾ cup small dry pasta shells
1 cup diced tomatoes
Salt and pepper to taste

In a large saucepan heat the olive oil. Add the onion, mushrooms, basil, oregano, and marjoram; sauté. Cook until the vegetables are slightly browned, about 10 minutes. Stir in the broth, squash, and potato. Bring to a boil, then reduce heat; cover and boil about 5 minutes.

Add the pasta and cook, covered, about 7 to 8 minutes, until the pasta is tender to bite. (Don't overcook pasta.) Stir in the tomatoes and heat for about 2 minutes. Season with salt and pepper.

Makes 6 servings.
193 calories per serving when made with Quick Garlic Soup.
220 calories per serving when made with chicken or vegetable stock.

Asparagus Soup

2 tablespoons olive oil
1 small onion, chopped
3½ cups chopped asparagus tips
  (about 1 pound)
4 cups chicken or vegetable stock
  or Quick Garlic Soup
  (see page 185)

1 cup spinach, chopped
2 teaspoons fresh tarragon,
  chopped
⅛ teaspoon nutmeg
salt and pepper
  to taste

In a medium saucepan, heat the olive oil. Add the onion and sauté until it is translucent, about 6 minutes. Stir in asparagus and cook, stirring, for 1 minute more.

Add the stock and bring to a boil. Cover and simmer until onion and asparagus are tender, about 30 minutes.

Transfer contents to a food processor or blender. Add the spinach and tarragon, and puree until smooth. Add nutmeg. Return the mixture to the pan; add salt and pepper to taste, and reheat for about 3 minutes.

Sprinkle nutmeg on each bowlful before serving.

Makes 6 servings.
81 calories per serving when made with Quick Garlic Soup.
108 calories per serving when made with chicken or vegetable stock.

Black Bean and Crab Bisque

1 tablespoon olive oil	pinch freshly ground black pepper
1 garlic clove, minced or pressed	6 ounces cooked crab meat,
½ cup onion, minced	fresh or canned
3 cups chicken stock or Quick	2 tablespoons brandy
Garlic Soup (see page 185)	fresh chives, chopped (optional)
1 pound black beans, cooked	

In a medium saucepan, heat the olive oil. Add the garlic and cook until fragrant. Add the onion and cook until translucent, about 10 minutes. Add the stock, beans, and pepper. Reduce the heat and simmer, uncovered, for 30 minutes.

Pour the mixture into a blender container. Blend on high until smooth. Pour mixture back into the saucepan, add the crab meat and brandy, and mix well. Heat through. Garnish with chopped chives.

Makes 4 servings.
263 calories per serving when using Quick Garlic Soup.
290 calories per serving when using chicken stock.

Yellow Split Pea Soup

7 cups water	2 to 4 garlic cloves, minced or
2 cups dried yellow split peas	pressed
1 large onion, chopped	2 tablespoons vegetable oil
2 tomatoes, peeled and chopped	2 tablespoons tamari or soy sauce
2 stalks celery, sliced	1 sweet potato, diced

Place water, split peas, onion, tomatoes, celery, garlic, oil, and tamari in a large, heavy-bottomed saucepan. Bring to a boil.

Reduce heat, partially cover, and simmer for 30 minutes. Add the sweet potato; cover and simmer another 30 minutes.

Makes 8 servings.
232 calories per serving.

Wheat Berry Soup

*Wheat berries sound weird to most of us. In fact, very few people
have tried them. Wheat when processed is not a Good Calorie;
however, wheat berries have not been processed at all, so they are an
exceptionally Good Calorie. They are delicious and have a wonderful
crunchy texture. They can be purchased in health-food stores.
They can also be cooked for breakfast.*

1 cup wheat berries	4 cups chicken or vegetable
1 cup navy beans	stock, Quick Garlic Soup
2 tablespoons olive oil	(see page 185), or water
1 onion, chopped	1 tablespoon dried thyme
1 leek, chopped	1 teaspoon dried rosemary
1 celery stalk, chopped	1 bay leaf
2 to 3 garlic cloves, minced	salt and pepper to taste

Soak the wheat berries and navy beans separately, preferably overnight. (Three hours of soaking is the minimum.) Drain.

In a large soup pot, heat the olive oil. Add the onion, leek, celery,
and garlic. Sauté until the onion is translucent, about 10 minutes.
Add the soaked navy beans, broth, and seasonings. Bring to a boil,
then reduce heat to a simmer and cook until beans are done, about
an hour.

Makes 6 servings.
257 calories per serving when using Quick Garlic Soup.
285 calories per serving when using chicken or vegetable stock.

SALAD DRESSINGS

Garlic and Buttermilk Dressing

½ cup low-fat buttermilk
3 scallions, thinly sliced
1 teaspoon seasoned rice vinegar
   or 1 tablespoon white wine
   vinegar plus ½ teaspoon sugar

2 teaspoons Dijon mustard
1 garlic clove, minced or pressed
salt and pepper to taste

Mix the buttermilk, scallions, vinegar, mustard, and garlic together.
Add salt and pepper to taste.

Makes about ⅔ cup.
10 calories per tablespoon.

Creamy Thousand Island Dressing

This dressing can be used over salads or as a sandwich spread.

¾ cup plain nonfat yogurt
3 tablespoons dry nonfat milk
   powder or nonfat milk
¼ cup unsalted and unsweetened
   tomato puree

3 tablespoons frozen unsweetened
   apple juice concentrate or
   apple butter
2 tablespoons red wine vinegar
2 tablespoons low-fat mayonnaise

Whisk the yogurt and milk powder together. Add all the other ingredients and stir until smooth.

Makes 1½ cups.
14 calories per tablespoon.

Creamy Low-Fat Dressing

¼ cup part-skim ricotta cheese
¼ cup plain nonfat yogurt
1 small garlic clove, minced or
    pressed
1 tablespoon minced fresh chives
    or fresh basil

1 tablespoon fresh parsley,
    minced
salt and pepper to taste

Blend the ricotta, yogurt, and garlic in a food processor or blender.
Then stir in chives and parsley. Add salt and pepper to taste.

Makes ½ cup.
10 calories per tablespoon.

Low-Fat Blue Cheese Dressing and Dip

1 cup low-fat cottage cheese
2 tablespoons blue cheese, crumbled

½ cup plain nonfat yogurt
garlic powder to taste

Blend all ingredients in a blender, food processor, or by hand.
    Serve with raw vegetables or as a salad dressing.

Makes 1½ cups.
13 calories per tablespoon.

Tamari-Ginger Sauce or Dressing

This is a great sauce for steamed vegetables and rice.

¼ cup water
¼ cup tamari

¼ to ½ teaspoon fresh grated
    ginger

Mix all ingredients. Spoon over vegetables or rice.

Makes ½ cup.
35 calories per ½ cup.

Yogurt Dressing

Basic Dressing:
1 cup nonfat yogurt
½ cup lemon juice
parsley

onions, minced
scallions, minced

Optional: 1 tablespoon Roquefort cheese

Mix basic ingredients.

Makes ½ cup.
6 calories per tablespoon.
13 calories per tablespoon with cheese.

Cottage Cheese Dressing

Basic Dressing:
1 cup low-fat cottage cheese
½ cup lemon juice

1 garlic clove, pressed
vegetable seasoning

Optional: 1 tablespoon Parmesan cheese

Mix basic ingredients, either by hand or in a blender, and add Parmesan cheese, if desired.

Makes 1½ cups.
10 calories per tablespoon.
11 calories per tablespoon with cheese.

SALADS

Cucumber, Tomato, and Yogurt Salad

3 cups nonfat plain yogurt
2 small ripe tomatoes, seeded and
   chopped
1 small cucumber, peeled, seeded,
   and finely diced or shredded
1 small onion, minced

1 fresh green chili pepper, seeded
   and finely sliced (optional)
3 tablespoons fresh cilantro,
   chopped
salt to taste (optional)

Place the yogurt in a bowl and whisk with a fork until creamy and smooth. Add tomatoes, cucumber, onion, chili, cilantro, and salt.

Makes 4 servings.
111 calories per serving.

Peach, Apple, and Yogurt Salad

Any fruit can be substituted in this salad.

1½ cups plain nonfat yogurt
2 fresh peaches, cut into
   1-inch pieces (1 cup)

1 fresh apple, sliced

Drain the yogurt in a sieve lined with cheesecloth for 20 minutes. Discard liquid.

Combine all the ingredients in a bowl. Cover and refrigerate until ready to serve.

Makes 4 servings.
85 calories per serving.

Vegetable–Cottage Cheese Salad

2 cups low-fat cottage cheese
½ cup scallions, thinly sliced
½ cup radish, finely chopped
⅓ cup celery, finely chopped
¼ cup red or green bell peppers,
   finely chopped
¼ cup shredded carrots

2 tablespoons fresh parsley,
   chopped
1 teaspoon fresh dill, chopped,
   or ½ teaspoon dry dill
½ teaspoon garlic salt
½ teaspoon dry mustard
paprika

Combine all the ingredients except paprika. Mix lightly. Cover and refrigerate.

Sprinkle with paprika before serving.

Makes 4 servings.
97 calories per serving.

Roasted Vegetable and Pasta Salad

2 eggplants, cut into ¼-inch rounds
2 tablespoons salt
1 cup dry pasta
olive oil spray
4 small zucchinis, cut into
   ¼-inch rounds
1 red bell pepper
2 scallions, chopped

2 tablespoons fresh parsley,
   chopped

Dressing:
½ cup plain nonfat yogurt
1 garlic clove, minced or pressed
2 tablespoons fresh basil or
   1 tablespoon dry basil

Preheat oven to 350°.

Put the eggplants in a colander. Sprinkle with the salt and toss to coat. Let sit for 30 minutes. Rinse eggplant and pat dry. (This will remove any bitterness in the eggplant.)

Meanwhile, cook the pasta until al dente. Run under cold water.

Spray a baking sheet with the olive oil spray. Arrange eggplant and zucchini slices in a single layer. Spray with olive oil spray twice. Bake until golden brown, about 30 minutes, turning slices every 10 minutes. Remove from the oven.

Turn the broiler on. Line the broiler rack with aluminum foil.

Cut the red pepper in half lengthwise; core and seed.

Place the red bell pepper on the rack, cut side down, about 3 inches from broiler. Broil until charred, about 5 minutes. Remove and gently place in zipper-closure plastic bag. Allow to cool for 10 minutes. Then, remove from plastic bag, rub off skin, and discard; chop remaining pulp.

Mix the eggplants, zucchinis, roasted red pepper, pasta, scallions, and parsley in a large bowl.

Combine all the ingredients for the dressing in a small bowl.

Pour the dressing over pasta and roasted vegetables and toss well.

Makes 4 servings.
160 calories per serving.

Garden Pasta Salad

1 pound corkscrew pasta
½ cup red onion, diced
1 zucchini squash, finely diced
6 radishes, trimmed and diced
2 large ripe tomatoes, cored,
    seeded, and finely diced

1 green bell pepper, finely diced
1 red bell pepper, finely diced
¼ cup fresh parsley, finely
    chopped

Cook pasta only until al dente. Run under cold water.

Mix all vegetables in a large bowl.

Toss salad with dressing of choice.

Makes 8 servings.
228 calories per serving without dressing.

Oriental Vegetable Salad

1 pound fresh asparagus
1 ounce cellophane noodles,
 uncooked
7 dried porcini mushrooms
1½ cups bean sprouts
½ medium carrot, cut into
 julienne strips
¼ cup water chestnuts, chopped
¼ teaspoon sugar

1 teaspoon dry mustard
¼ teaspoon salt (optional)
3 tablespoons rice vinegar or
 white wine vinegar
1½ tablespoons tamari or soy
 sauce
2 teaspoons sesame oil
 (dark preferred)
8 romaine lettuce leaves

Snap off tough ends of the asparagus. Cut asparagus into 1-inch pieces. Blanche in boiling water for 30 seconds. Drain; rinse under cold water until cool. Set aside.

Combine noodles and mushrooms in boiling water to cover. Let stand until soft, about 15 minutes. Drain well. Slice mushrooms into thin strips. Combine noodles, mushrooms, asparagus, bean sprouts, carrot, and water chestnuts; toss well. Cover vegetable mixture and chill thoroughly.

Meanwhile, combine sugar, mustard, salt, vinegar, tamari, and sesame oil; stir well. Pour over vegetable mixture and toss gently. Serve on bed of lettuce leaves.

Makes 4 servings.
95 calories per serving.

Rice Salad with Vegetables and Herbs

3 cups hot cooked brown or
 white rice
2 tablespoons lemon juice
2 tablespoons olive oil
1 garlic clove, minced or
 pressed
½ teaspoon dried oregano
½ teaspoon dried rosemary
½ teaspoon black pepper

½ teaspoon salt
1 small zucchini, julienned or
 grated
1 small yellow squash, julienned
 or grated
1 medium tomato, seeded and
 chopped
1 tablespoon grated Parmesan
 cheese

Place the rice in a large bowl.

Combine the lemon juice, olive oil, garlic, oregano, rosemary, pepper, and salt in a small jar with a lid. Shake well and pour over rice. Toss lightly.

Cover and let cool. Stir in zucchini, squash, tomato, and Parmesan cheese. Adjust seasonings.

Makes 6 servings.
105 calories per serving.

Sesame Green Bean Salad

1 pound fresh green beans
1 tablespoon olive oil
1 to 2 garlic cloves, minced or
    pressed
½ teaspoon crushed red pepper

3 tablespoons lemon juice
salt and pepper to taste
1 tablespoon sesame seeds,
    toasted

Wash the beans; trim ends and remove strings. Cook, uncovered, in boiling water until crisp-tender, about 5 minutes. Drain; rinse under cold water, and drain again.

Heat the olive oil in a small nonstick skillet. Add the garlic and red pepper; sauté until garlic is browned. Remove from heat and let cool.

Combine garlic mixture, lemon juice, and salt and pepper to taste; stir well. Combine beans, lemon juice mixture, and sesame seeds in a medium-size bowl; toss well. Cover and chill for at least 1 hour.

Makes 4 servings.
61 calories per serving.

Cauliflower and Broccoli Tossed Salad

2½ cups fresh cauliflower
    florets
2½ cups fresh broccoli florets
5 cups Bibb lettuce, torn
1 cup red bell pepper, chopped
¼ cup chicken or vegetable broth

¼ cup water
2 tablespoons lemon juice
1 tablespoon olive oil
1 garlic clove, minced or pressed
1½ teaspoons crushed red pepper
salt and pepper to taste

Steam the cauliflower and broccoli florets until crisp-tender, about 4 to 5 minutes. Transfer the vegetables to a large bowl. Add the lettuce and red bell pepper; toss gently.

Combine the chicken broth, water, lemon juice, olive oil, garlic, crushed red pepper, salt, and black pepper in a small jar. Cover tightly, shake, and blend well. Pour over vegetable mixture; toss gently. Serve immediately.

Makes 4 servings.
74 calories per serving.

Chickpea Salad with Tomatoes and Vinaigrette

4 ripe tomatoes
1 onion, chopped
2 to 3 garlic cloves, pressed or
    crushed
2 tablespoons olive oil
1 to 2 tablespoons red wine vinegar
2 teaspoons fresh rosemary,
    chopped
1 teaspoon paprika
pinch cayenne pepper

pinch sugar (optional)
salt and pepper, to taste
1½ cups chickpeas
    (garbanzo beans), cooked
1½ cups green bell pepper,
    coarsely chopped
½ cup red bell pepper,
    coarsely chopped
1 tablespoon fresh parsley,
    chopped

For the dressing: Puree two of the tomatoes, half of the onion, and the garlic cloves in a food processor or blender. Add the olive oil, vinegar, rosemary, paprika, cayenne, and sugar (if the tomatoes are acidic).

For the salad: Chop the remaining tomatoes and onion; combine them with the dressing. Season with salt and pepper. Mix chickpeas with bell peppers, then toss with dressing. Garnish with fresh parsley.

Makes 4 servings.
222 calories per serving.

Chicken Salad

2 cups boiled or leftover cooked
  chicken, cut into strips
1 small onion, thinly sliced
1 medium tomato, chopped
2 tablespoons Italian parsley
  or cilantro

½ cup red bell pepper, julienned
salt and pepper to taste
1 tablespoon olive oil
juice of ½ lemon

Mix all ingredients together except the olive oil and lemon juice. Then, in a separate bowl, mix the olive oil and lemon juice; stir, then pour over the chicken mixture.

Makes 2 servings.
229 calories per serving.

Basil-Lime-Chicken Salad

4 cups chicken stock
½ cup chopped fresh basil
¼ cup fresh lime juice
freshly ground pepper, to taste

pinch salt
4 skinless, boneless chicken
  breasts, cut in half

Combine the chicken stock, basil, lime juice, pepper, and salt; pour into a frying pan large enough to hold the chicken breasts in 1 layer. Bring liquid to a boil.

Layer chicken in liquid without overlapping; cover and simmer about 4 minutes. Remove from heat and allow chicken to cool at room temperature. Refrigerate overnight. Slice and toss in a green salad with vinaigrette.

Makes 4 servings.
176 calories per serving (without dressing).

PASTA

In order for pasta to remain a Good Calorie, it is essential not to overcook it. All pasta should be firm, al dente style, not soft and mushy. Note: Although processed wheat is a Bad Calorie, whole wheat and white pasta are Good Calories.

Spinach Fettuccine with Sun-dried Tomatoes

10 sun-dried tomatoes
1 pound spinach fettuccine
2 tablespoons olive oil
1 to 2 garlic cloves, minced or pressed
1 medium onion, chopped

½ cup red bell pepper, sliced
½ cup mushrooms, sliced
1 cup spinach, coarsely chopped
½ teaspoon ground nutmeg
salt and pepper to taste

Place the sun-dried tomatoes in a small bowl; pour boiling water over tomatoes to cover. Let stand for about 15 minutes or until tomatoes are tender. Drain tomatoes and discard liquid. Cut tomatoes into strips.

Meanwhile, cook the pasta. Drain.

Heat the olive oil in a large saucepan. Add garlic, onion, and red bell pepper; sauté until vegetables are tender, yet crisp, about 3 minutes. Add mushrooms and spinach; stir for 1 minute. Add sun-dried tomatoes, pasta, nutmeg, salt, and pepper. Cook and stir for 1 to 2 minutes or until heated through.

Toss vegetables over pasta.

Makes about 6 servings.
336 calories per serving.

Fresh Basil and Tomato Pasta

1 pound spaghetti, white or
   whole wheat
2 pounds ripe tomatoes, peeled
2 tablespoons olive oil
5 to 7 garlic cloves, minced
   or pressed

12 large fresh basil leaves, cut
   into slivers
½ teaspoon red pepper flakes
salt to taste
pinch nutmeg

Begin cooking the spaghetti.

Meanwhile, cut the tomatoes in half, gently scoop out seeds, and set aside.

Heat the olive oil in a medium saucepan; add garlic and sauté garlic for 1 minute. Add tomatoes, basil, red pepper, and salt to taste. Cook over medium heat for 5 minutes while crushing tomatoes with the back of a slotted spoon. Add nutmeg.

Toss tomato sauce with spaghetti.

Makes 4 servings.
351 calories per serving.

Pasta with Fresh Vegetables

This is a favorite recipe made by Margrit,
Monika's mother. It always gets raves.

1 pound rotini pasta or ziti
4 to 6 garlic cloves, minced
½ onion, chopped
2 tablespoons olive oil
½ cup fresh parsley
1 cup small broccoli florets
1 cup fresh or frozen peas

1 cup red bell pepper, chopped
1 cup green bell pepper, chopped
1 small zucchini, diced
2 medium red ripe tomatoes,
   chopped
12 fresh basil leaves, cut in slivers,
   or 1 teaspoon dried basil

Begin cooking the pasta.

Meanwhile, in a large nonstick skillet, sauté the garlic and onion in olive oil until onion is translucent. Add the parsley and cook just until it wilts. Stir in the broccoli, peas, and bell peppers and sauté for 5 to 7 minutes. Add the zucchini and cook for another 5 minutes. Add the tomatoes and basil and cook for about 3 to 4 minutes. Remove from heat. The tomatoes should be chunky and the vegetables should still retain their fresh-looking color.

Serve over pasta.

Makes 6 servings.
354 calories per serving.

Pasta with Asparagus and Roasted Peppers

2 large red bell peppers
3 tablespoons olive oil
4 large garlic cloves, minced or pressed
2 tablespoons balsamic vinegar
1 cup fresh basil leaves, slivered
Salt and pepper to taste

1 pound ripe tomatoes, peeled, seeded, and diced
1 cup fresh peas, shelled, or frozen peas, thawed
½ pound asparagus, trimmed and cut into ½-inch lengths
1 pound fusilli pasta
½ cup fresh parsley, chopped

Preheat over to 400°. Bake the whole bell peppers on baking sheet for about 45 minutes, turning every 10 minutes, until skins are puffy and brown. Transfer peppers to a bowl and cover with plate, and let sit for at least 30 minutes.

Remove the skins and seeds from the peppers, holding them over a bowl so as not to lose any of the juices. Cut peppers into thin strips; cut strips in half crosswise, and combine with the liquid in the bowl. Mix in the olive oil, garlic, vinegar, ½ cup of the basil, and the salt and pepper. Add the tomatoes. Cover and marinate for 1 hour. (Can be marinated overnight.)

Steam peas and asparagus until tender, about 5 minutes. Run vegetables briefly under water to cool. Set aside.

Cook pasta al dente. Drain and toss with marinade, vegetables, parsley, and remaining basil leaves. Season with salt and pepper to taste.

Makes 6 servings.
338 calories per serving.

Spinach Lasagna

½ pound lasagna noodles
2 pounds fresh spinach
1 cup part-skim ricotta cheese
4 tablespoons Parmesan cheese, grated
¼ teaspoon nutmeg
¼ teaspoon salt
freshly ground pepper, to taste
2 tablespoons olive oil

3 garlic cloves, minced or pressed
1 small onion, chopped
½ cup chopped bell pepper (red or green)
2 cups tomato sauce (canned or bottled)
1 teaspoon dried basil
1 teaspoon dried oregano
½ teaspoon dried thyme

Preheat oven to 350°.

Cook lasagna noodles. When they are done, rinse under cold running water.

Meanwhile, cut off the ends of the spinach and wash the leaves very well. Steam spinach until just limp, about 2 minutes. Chop the spinach; add the ricotta, 2 tablespoons of the Parmesan cheese, the nutmeg, salt, and ground pepper.

Heat the oil in a large saucepan. Add the garlic, onion, and bell pepper, and sauté until onion is translucent. Stir in the tomato sauce, basil, oregano, and thyme. Cover and let simmer for 3 to 5 minutes.

In a 8 × 13-inch baking dish layer the noodles first, then alternate with spinach-cheese mixture, then tomato sauce. Sprinkle the remaining Parmesan cheese on top.

Bake until bubbly, about 30 minutes.

Makes 8 servings.
236 calories per serving.

Pad Thai

This is a lower-fat version of the traditional Thai pasta dish.

3 tablespoons fish sauce*
2 tablespoons oyster sauce*
½ teaspoon sugar
1 stalk lemongrass*, finely chopped
2 cloves garlic, minced
⅛ to ¼ teaspoon crushed red pepper
1 package extra-firm low-fat tofu, cubed
1 teaspoon cornstarch

1 tablespoon sesame oil
3 scallions, sliced diagonally into ½-inch slices
1 cup coarsely shredded bok choy or Napa cabbage
½ cup red bell pepper, julienned
2 cups rice noodles, cooked
2 cups bean sprouts

In a medium bowl, mix the fish sauce, oyster sauce, sugar, lemon-grass, garlic, and crushed red pepper. Add the tofu. Marinate in the refrigerator for up to 8 hours.

Preheat wok or large sauté pan over medium-high heat.

Meanwhile, drain marinade from tofu mixture into a small bowl. In another small bowl, dissolve cornstarch with a little water. Stir cornstarch mixture into marinade. Add sesame oil and marinade to wok. Sauté scallions, cabbage, red bell pepper, and tofu for 4 to 5 minutes. Add cooked rice noodles, mixing enough to allow mari-nade to coat noodles; then add bean sprouts. Cook for 30 seconds more. Remove from heat.

Makes 4 servings.
281 calories per serving.

*These items can be purchased at an Asian grocery store or health-food store.

GRAINS

Bulgur

Bulgur can be used in place of rice, served as a side dish, or used in stirfries or as a stuffing when mixed with herbs and spices. Bulgur is also called cracked wheat.

2 cups water
1 cup bulgur

salt to taste (optional)

Boil water in a medium saucepan, then add bulgur and salt. Reduce heat and simmer for 5 minutes. Remove from heat. Let sit, covered, for 10 minutes.

Or, boil the water, pour over the bulgur, and remove it from the heat. Let sit for 1 hour.

Makes approximately 3 cups.
152 calories per cup.

Tabouli

1 cup bulgur
2 cups water
1 garlic clove, minced or pressed
juice of 4 large lemons
salt and pepper to taste

8 cups parsley, finely chopped
1 large bunch fresh mint, finely
  chopped
4 scallions, thinly sliced
3 ripe tomatoes, diced

Prepare bulgur as described above. Allow bulgur to cool.

In a large bowl, toss with garlic, lemon juice, and salt and pepper to taste. Then toss with parsley, mint, scallions, and ripe tomatoes. Add more lemon juice, salt, and pepper, if desired.

This recipe can be varied by adding more or less bulgur.

Makes about 6 cups.
136 calories per cup.

Lemon-Bulgur Timbales

2 tablespoons vegetable oil
4 tablespoons shallots,
   thinly sliced
1 cup mushrooms, minced
1 cup coarse bulgur
1 cup stock (any kind)

1 tablespoon lemon rind,
   freshly grated
5 tablespoons fresh chives,
   finely chopped
salt and pepper to taste
lemon slices for garnish

Heat the vegetable oil in a medium skillet. Add the shallots; sauté until transparent. Add the mushrooms and sauté for an additional few minutes.

Stir in the bulgur; add the stock and bring to a boil. Cover and simmer until bulgur is cooked, about 10 minutes.

Fluff the bulgur lightly. Add the lemon rind and chives. Remove from heat, cover, and let stand for 5 minutes.

Season with salt and freshly ground pepper.

Pack bulgur into ½-cup oiled molds and leave for 10 minutes. Then invert molds and garnish with lemon slices.

Makes 4 servings.
150 calories per serving.

Vegetable Rice Pilaf

2 tablespoons vegetable oil
1 onion, chopped
2-inch cinnamon stick
½-inch fresh ginger,
   finely chopped
¼ teaspoon cayenne pepper
2 teaspoons cumin seeds
2 cups water

1 cup rice
1½ cups fresh mixed vegetables
   (green beans, peas, zucchini,
   etc.—can use frozen
   vegetables)
½ teaspoon turmeric
¼ teaspoon salt

Heat the vegetable oil in a medium saucepan. Add the onion, cinnamon stick, ginger, cayenne, and cumin seeds; sauté until onion is translucent and very soft, about 8 minutes.

Add the water and bring to a boil. Add the rice, vegetables, turmeric, and salt. Lower heat, cover, and simmer until rice is done, approximately 15 to 20 minutes.

Makes 6 servings.
156 calories per serving.

Margrit's Curried Brown Rice

White rice can be substituted for the brown; just adjust cooking time.
The leftover rice can be used as a base for a delicious salad.
Just add some raw vegetables and a few lettuce leaves,
and toss with a little vinaigrette.

2 tablespoons olive oil	3 cups water
2 to 3 garlic cloves, minced	1 teaspoon curry powder
1 onion, chopped	1 chicken bouillon cube
3 cups brown rice	salt and pepper to taste

Heat the olive oil in a medium saucepan. Add the garlic and onion; sauté until golden brown, about 6 to 7 minutes. Add the rice and mix well, coating all the grains. Add the water, curry, bouillon cube, salt, and pepper, stirring constantly until it boils. Cover and simmer for 35 minutes, fluffing lightly a few times with a large-pronged cooking fork. (This keeps the rice kernels nice and chewy.)

Uncover, fluff lightly, and simmer for 5 additional minutes.

Makes 6 cups.
195 calories per ½ cup.

Wheat Berries

1 cup wheat berries	3 cups water

Soak wheat berries overnight.

Bring the water to boil. Add wheat berries. Reduce heat and simmer for 1 to 1½ hours.

Makes 4 servings.
157 calories per serving.

Lemon Rice

2 tablespoons olive oil
1 onion, finely chopped
3 cups white rice
2 cups chicken stock
1 cup water
juice of 3 lemons

5 lemon slices
1 teaspoon sea salt
freshly ground pepper to taste
2 bay leaves
4 whole cloves

Heat the olive oil in a medium saucepan. Add the onion; sauté until translucent, about 3 minutes. Stir in the rice, coating the grains with the oil, and stir for 4 to 5 minutes.

Meanwhile, in a small saucepan, combine the chicken stock, water, lemon juice, lemon slices, salt, pepper, bay leaves, and cloves, and bring to a boil. Stir this mixture into the rice; cover and simmer over low heat until liquid is absorbed, about 20 to 25 minutes.

Makes 8 servings.
310 calories per serving.

Barley

1 cup barley
5 cups water
½ teaspoon salt (optional)

2 tablespoons fresh parsley,
   chopped (optional)
½ to 1 teaspoon cumin (optional)

Put barley, water, and salt in a medium saucepan and bring to a boil. Cover and reduce heat to low until barley is tender, about 55 to 60 minutes. Or place ingredients in a pressure cooker, bring to a boil, lock cover, and cook over medium heat for about 45 minutes.

Add parsley and cumin if desired.

Makes approximately 3 cups.
193 calories per cup.

Barley with Mushrooms

2 tablespoons vegetable oil
4 to 5 garlic cloves, minced or pressed
1 cup fresh parsley, chopped
1 cup mushrooms, sliced

3 tablespoons dry white wine
1 cup barley
2 cups chicken or vegetable
   broth

Heat the oil in a medium saucepan. Add garlic and parsley; sauté until garlic is fragrant and parsley is wilted.

Add the mushrooms. Cook, stirring constantly, until they darken, about 3 minutes. Add wine and barley and continue cooking for another few minutes.

Add broth, bring to a boil, then lower to medium heat. Cover and simmer for 40 minutes. Remove from heat and let stand, uncovered, for 10 minutes. ´

Makes 6 servings.
145 calories per serving.

LEGUMES

Everything You Always Wanted to Know About Beans

All legumes are Good Calories. Not only are they an excellent source of protein and fiber, they are inexpensive as well. Unfortunately, people often shy away from making them because they have no experience cooking them. Don't let this stop you. Beans are incredibly easy to make.

It is important to soak them (except for split peas and lentils) overnight or for at least 4 hours; then discard the soaking water (this makes them easier to digest, causing less gas). Then cook according to times listed on page 212. For even faster cooking, use a pressure cooker. They can also be bottled and stored, or frozen.

All the following recipes for beans require that they be washed and soaked (if soaking is required for the specific bean). Canned beans can be substituted; however, dried beans have a lower glycemic index and therefore reduce insulin better.

Quick Soaking Method

Cover beans with cold water and bring to a boil, then simmer for a few minutes. Remove from heat, cover, and let stand for 1 to 1½ hours. Beans are usually ready to cook.

Overnight Soaking Method

Soaking dried beans overnight is best; it makes the beans easier to digest, preventing bloating and gas.

Beans That Require Soaking

Type of Bean	Regular Cooking Time	In a Pressure Cooker	Calories per ½ Cup (Cooked)
adzuki	2 hours	30 minutes	147
black beans	1½ hours	30 minutes	113
black-eyed peas	1 to 1¼ hours	25 minutes	92
chickpeas (garbanzos)	3 hours	35–40 minutes	134
fava beans	1½ to 2 hours	16–18 minutes	90
great northern beans	2 hours	30 minutes	104
lima beans	1½ hours	not recommended	104
navy or pea beans	2½ hours	20–30 minutes	129
pinto beans	2½ hours	30 minutes	117
red kidney beans	1½ hours	25 minutes	112
soybeans	3 hours	not recommended	149

Beans That Do Not Require Soaking

(Can be soaked to make them even more digestible)

Type of Bean	Regular Cooking Time	In a Pressure Cooker	Calories per ½ Cup (Cooked)
brown lentils	1¼ hours	20 minutes	115
green split peas	1¼ hours	25 minutes	116
red lentils	20–25 minutes	not recommended	115
yellow split peas	1¼ hours	25 minutes	116

Helpful Hints:

♦ 1 pound of dry beans is approximately 2½ to 3 cups, which will yield 6 to 7 cups of cooked beans.

♦ Cooked beans freeze well.

♦ Cooked beans can also be stored in airtight glass containers (away from heat) for up to 6 months. Include the cooking liquid.

♦ Beans that are cooked with additional items such as tomatoes, onions, and oils take a little longer to cook.

Basic Bean Recipe

1 cup dried beans

1 or 2 bay leaves*

4 cups cold water

Soak the beans in the cold water for at least 4 hours (in refrigerator if kitchen is warm). Remove any beans that float. Change water. Add bay leaf and bring beans to a boil and simmer according to times listed on page 212. Add salt only after beans are tender.

*The bay leaf helps to make beans easier to digest.

Black Bean Chili

1 pound dried black beans
2 tablespoons olive oil
3 medium onions,
  chopped
4 to 6 large garlic cloves,
  minced or pressed
6 cups water
1 bay leaf
One 28-ounce can tomatoes,
  seeded and chopped, with
  the juice

1 ounce dried tomatoes,
  chopped (optional)
4 teaspoons cumin
1 tablespoon paprika
½ to 1 teaspoon cayenne pepper
2 tablespoons chili powder
2 teaspoons dried oregano
1 green bell pepper, diced
salt and pepper to taste
1 tablespoon cider vinegar
½ cup fresh cilantro

Soak beans according to directions in Basic Bean Recipe (above). Heat olive oil in a large stockpot; add onions and garlic. Sauté until onions are translucent. Add the soaked beans, water, and bay leaf. Bring to a boil. Reduce heat, cover, and simmer for 1 hour.

Add the tomatoes, spices, and green bell pepper. Cover and simmer for 1½ hours, or until beans are tender. Remove bay leaf. Stir in vinegar and cilantro just before serving.

Makes 6 servings.
285 calories per serving.

Lentil and Barley Pilaf

2 tablespoons olive oil
2 onions, chopped
4 to 5 garlic cloves, minced
1 teaspoon cumin seeds
1 tablespoon paprika
½ teaspoon curry powder

pinch ginger
l cup cooked lentils
1 cup cooked barley
2 bay leaves
1 cup of any stock
salt and pepper to taste

Heat the olive oil in a medium saucepan. Add onions; sauté until they turn golden brown. Add the garlic and cumin and cook a few minutes. Add paprika, curry, and ginger, and cook a few more minutes. Add lentils, barley, and bay leaves, and stir.

Add half of the stock. Continue to cook and stir until liquid has been absorbed. Add the remaining stock and continue to cook until all the liquid has been absorbed. Add salt and pepper to taste.

Makes 4 servings.
190 calories per serving.

Middle Eastern Lentils and Rice

2 tablespoons olive oil
2 large onions, finely chopped
4 garlic cloves, minced or pressed
¾ teaspoon salt
¾ teaspoon ground allspice
¾ teaspoon ground cinnamon
2 cups dried lentils

½ cup brown or white rice
4½ cups water
2 large onions,
    thinly sliced vertically
3 cups plain nonfat yogurt
    (optional)

Heat 1 tablespoon of the olive oil in a large stockpot. Add half of the onions, the garlic, salt, allspice, and cinnamon, and sauté, until onions are translucent.

Add the lentils and rice, and continue to cook, stirring, for 5 more minutes. Add the water and bring to a boil. Cover and cook over low heat until the liquid is absorbed and the rice is tender, about 1 hour for brown rice or 20 to 30 minutes for white rice.

Meanwhile, heat the remaining tablespoon of olive oil in a large skillet, and sauté the remaining onions, stirring frequently, until the onions are very brown but not scorched.

To serve, spoon lentil and rice mixture onto a serving platter and cover with browned onions. Serve with yogurt.

Makes 6 servings.
449 calories.

Creamy Rice and Beans

1 cup black beans
1 cup white beans
6 cups water or vegetable
 or chicken stock
2 onions, chopped

5 cloves garlic, minced or pressed
1 tablespoon olive oil
1 teaspoon salt
¼ cup dried lentils
1 cup brown rice

Soak black and white beans according to directions in Basic Bean Recipe (page 213).

Place black beans, white beans, and water in a large stockpot and bring to a boil. Reduce heat and add onions, garlic, olive oil, and salt. Simmer for about 3½ hours. Stir occasionally.

Add lentils and rice and bring to a boil. Reduce heat and simmer until rice and lentils are tender and the broth is creamy, about 40 minutes. Make sure that rice does not overcook.

Makes about 12 servings.
201 calories per cup.

Creamy Kidney Beans and Lentils

½ cup dried red kidney beans
   or pinto beans
½ cup dried lentils
2½ cups water
3 teaspoons unsalted butter
½ cup onion, minced
1 to 2 garlic cloves, minced or
   pressed
1½ teaspoons fresh ginger, minced

2 fresh hot chili peppers,
   stemmed and minced
¼ cup unsweetened tomato paste
½ teaspoon ground coriander
½ teaspoon ground cumin
freshly ground pepper, to taste
½ teaspoon salt
½ cup low-fat milk
fresh cilantro, chopped, for garnish

Soak beans according to directions in Basic Bean Recipe (page 213). Place beans, lentils, and water in a large, heavy saucepan; bring to a boil. Reduce heat, cover, and cook until beans are tender, about 1½ hours. Remove from heat.

Heat butter in a large nonstick saucepan. Add onion, garlic, ginger, and chilies, and sauté until onion is lightly browned, about 5 minutes. Add tomato paste, coriander, cumin, and pepper, stirring constantly, for about 2 minutes. Add the cooked beans, salt, and milk. Bring to a boil; reduce heat to low. Cover and cook until most of the liquid is absorbed, about 7 to 9 minutes.

Garnish with fresh chopped cilantro.

Makes 4 servings.
255 calories per serving.

Escarole and Chickpeas

1 cup chickpeas
   (garbanzo beans), cooked
2 tablespoons olive oil
1 to 2 garlic cloves, minced
   or pressed (optional)
1 head of escarole, cut into
   ½-inch wide strips

¼ cup fresh parsley,
   chopped
1 teaspoon grated
   lemon zest
salt and pepper to taste
2 tablespoons
   lemon juice

Cook beans according to directions in Basic Bean Recipe (page 213).

Heat the olive oil in a large skillet. Add the garlic, and sauté just until garlic is fragrant. Add the escarole and continue stirring over medium heat until it is wilted. Add the parsley, lemon zest, salt, and pepper. Then add chickpeas and cook until they are warmed through. Sprinkle with lemon juice.

Makes 4 servings.
118 calories per serving.

Garlicky Pinto Beans

2 tablespoons olive oil
1 large onion, chopped
3 to 4 garlic cloves, minced
½ teaspoon ground cumin
¼ teaspoon dried oregano

4½ cups water
1 pound dried pinto beans
1 teaspoon salt
¼ teaspoon black pepper
¼ teaspoon red pepper

Heat the olive oil in a large stockpot. Add the onion, garlic, cumin, and oregano; sauté until onion is translucent. Add the water, beans, salt, black pepper, and red pepper. Bring to a boil, then reduce heat to low and simmer until beans are tender, about 1½ hours.

Makes 4 servings.
285 calories per serving.

Lentil and Spinach Stew

1 cup dried lentils
3 cups cold water
1 onion, chopped
salt to taste

red pepper flakes to taste
1 pound fresh spinach, washed
    and drained
2 tablespoons olive oil

Put lentils, water, and onion in a medium stockpot and bring to a boil; lower heat and simmer for 45 minutes.

Add the salt, red pepper flakes, and spinach; cover and boil gently for 6 to 7 minutes. Stir in the olive oil.

Makes 6 servings.
165 calories per serving.

VEGETABLES

Mashed Sweet Potatoes

5 large sweet potatoes, scrubbed
2 teaspoons butter
⅛ teaspoon nutmeg
⅛ teaspoon cinnamon

1 cup low-fat milk
1 teaspoon salt
freshly ground black pepper, to taste
1 to 2 teaspoons balsamic vinegar

Preheat oven to 400°. Bake the sweet potatoes for about 50 minutes, until easily pierced with a fork. Let potatoes sit until cool enough to handle. Peel and pass through a food mill, or mash by hand.

Heat the butter in a medium saucepan over a low flame until it browns. Stir in the nutmeg and cinnamon. Remove from heat and stir in the milk. Add the sweet potato puree and combine thoroughly. Season with salt and pepper, stir in balsamic vinegar.

Makes 4 servings.
278 calories per serving.

Sumptuous Broccoli

2 pounds broccoli, cut into florets
2 large tomatoes, coarsely chopped
¼ medium onion
2 garlic cloves
2 tablespoons olive oil
½ cup chicken or vegetable stock

1 bay leaf
½ teaspoon cumin
½ teaspoon salt
⅛ teaspoon cinnamon
freshly ground pepper, to taste
pinch sugar

Blanch the broccoli for 2 minutes; drain and set aside.

Puree the tomatoes, onion, and garlic in a food processor or blender until smooth.

Heat the olive oil in a medium skillet. Add the puree and cook for 1 minute. Add the stock and spices, and cook for 2 minutes, stirring occasionally.

Add broccoli, cover, and cook over medium-low heat until broccoli is tender-crisp, about 6 to 8 minutes. Discard the bay leaf before serving.

Makes 4 servings.
122 calories per serving.

Summer Squash Soufflé

1 cup summer squash, shredded
½ onion, minced
6 fresh mushrooms, sliced
2 teaspoons cornstarch
1 cup low-fat milk
2 eggs

½ teaspoon salt
¼ to ½ teaspoon pepper
¼ teaspoon nutmeg
⅓ cup low-fat mozzarella cheese, diced

Preheat oven to 350°.

In a large bowl, combine the summer squash, onion, and mushrooms.

In a medium bowl, mix together the cornstarch with a little of the milk, then add the remaining milk. Beat in the eggs and seasonings. Stir the egg mixture into the vegetables, then stir in the mozzarella. Pour into a 9-inch pie plate. Bake for about 45 minutes.

Makes 4 servings.
92 calories per serving.

Spaghetti Squash with Tomatoes

1 spaghetti squash
2 tablespoons olive oil
1 onion, chopped
1 cup tomatoes, preferably fresh
   (drain if canned)

1 tablespoon chopped fresh basil
   or 1 teaspoon dried
salt and pepper to taste

Cut squash in half lengthwise. Scoop out the seeds. Put the 2 halves in a large pot with a steamer and steam for 20 to 25 minutes. The spaghetti strands should be tender, yet firm.

Meanwhile, heat the olive oil in a medium saucepan and add the onion; sauté until it browns. Add the tomatoes, basil, salt, and pepper. Cover and simmer for 15 minutes. Run through a food mill or processor.

When the squash cools enough to handle, comb out the spaghetti strands. Mix the tomatoes into the strands.

Makes 4 servings.
140 calories per serving.

Sautéed Grated Zucchini

8 small zucchini (any variety: 2 onions, finely chopped
   green, yellow, sunburst, etc.)   freshly ground pepper to taste
1 tablespoon salt   Parmesan cheese to taste
2 tablespoons olive oil

Grate zucchini with the large holes of a cheese grater over a colander for drainage. Mix the salt into the zucchini and let sit for 20 minutes. Then squeeze grated zucchini in a cheesecloth to extract remaining juice.

Heat the olive oil in a large skillet and sauté the onions until they are translucent. Add the zucchini and pepper. Cook over moderate heat, stirring often, for about 5 minutes. Cover and let cook for an additional 8 minutes.

Sprinkle with Parmesan cheese.

Makes 6 servings.
82 calories per serving.

Spinach and Red Onion Stirfry

2 tablespoons rice vinegar   2 red onions, cut into rings
2 tablespoons tamari or soy sauce   1½ pounds spinach, washed,
1 teaspoon dark sesame oil   stemmed, and torn into small
¼ teaspoon sugar   pieces
1 tablespoon vegetable oil

In a small bowl whisk together the vinegar, tamari, sesame oil, and sugar. Set aside.

Heat the vegetable oil in a large skillet or wok. Add onion rings and stir over high heat for 45 seconds. Add spinach and stir for 45 seconds. Add tamari mixture and reduce heat. Stir continuously until the liquid is evaporated and the onions are crisp-tender.

Makes 4 servings.
90 calories per serving.

Curried Cauliflower

2 tablespoons vegetable oil
1 teaspoon cumin powder or seeds
1 teaspoon ground coriander
½ teaspoon ground turmeric
1 tablespoon fresh ginger,
   minced
½ teaspoon salt
pinch cayenne pepper
1 medium cauliflower, separated
   into small florets
½ to 1 cup water
1 large red ripe tomato,
   chopped (optional)
2 tablespoons fresh parsley

Heat the vegetable oil in a large skillet. Add cumin, coriander, and turmeric. Stir over medium heat until spices are aromatic. Be careful not to burn them. Stir in ginger, salt, and cayenne. Then add cauliflower, stirring to coat with spices. Add ½ cup of the water, cover pan, reduce heat, and simmer until the vegetables are crisp and tender, about 10 to 15 minutes. Add extra water only if needed.

Add tomato and parsley. Increase heat, and stir until the water is almost evaporated.

Can be served hot or cold.

Makes 4 servings.
90 calories per serving.

Vegetable Stirfry

2 tablespoons vegetable oil
1 to 2 garlic cloves, minced (optional)
¼ to ½ teaspoon fresh
  ginger, minced
2 cups vegetables, chopped in
  small pieces

soy sauce or tamari, to taste
2 tablespoons grated hard cheese
  (Parmesan, Romano, or white
  cheddar) (optional)

Heat the vegetable oil in a wok or large skillet over medium-high heat. Add the garlic and ginger, stirring for about 30 seconds. Do not let garlic brown.

Add vegetables that require the longest cooking time first, then add others that require less cooking. Add the soy sauce and a little water if necessary. Cook until vegetables are crisp-tender.

Remove from heat and sprinkle cheese on before serving. Serve over rice.

Makes 4 servings.
151 calories per serving.

Italian Frittata

2 tablespoons olive oil
1 cup fresh mushrooms, sliced
1 cup zucchini, sliced
½ red bell pepper, julienned
½ green bell pepper, julienned
1 scallion, chopped
1 garlic clove, minced or pressed
1 egg and 6 egg whites, lightly beaten

3 tablespoons freshly grated
  Parmesan cheese
freshly ground pepper to taste
dash dried basil
dash dried oregano
3 tablespoons water
dash paprika for garnish

Heat the olive oil in a large nonstick skillet. Add mushrooms, zucchini, peppers, scallion, and garlic; sauté, stirring occasionally, until vegetables are tender, about 5 minutes.

Meanwhile, in a medium-size bowl, combine eggs with 1 tablespoon of the Parmesan cheese. Add remaining seasonings and water. Pour mixture over cooked vegetables in pan. Leave pan cover slightly ajar, allowing steam to escape. Cook over medium heat until eggs are set, 6 to 8 minutes. Serve from skillet or loosen edges with a rubber spatula and slide onto serving plate.

Sprinkle with the remaining Parmesan and dash of paprika.

Makes 4 servings.
149 calories per serving.

Yam and Cheese Strudel

3 medium yams or sweet potatoes, peeled

4 tablespoons Parmesan cheese, grated

3 teaspoons fresh parsley or 1 teaspoon dried parsley

1 teaspoon toasted sesame seeds

salt and pepper to taste

2 tablespoons vegetable oil

Preheat oven to 425°.

Slice yams as thinly as possible.

Combine the dry ingredients in a small bowl.

Place vegetable oil in a small bowl.

Layer one-quarter of the yams in a casserole dish, then brush them lightly with the oil using a pastry brush. Sprinkle with 1 table-spoon of the Parmesan cheese mixture. Repeat until all the yams are used.

Bake covered for about 30 minutes. Uncover, and let strudel bake until browned on top, about 20 minutes.

Makes 6 servings.
150 calories per serving.

Vegetable-Saffron Paella

2 cups brown or white rice
6 cups vegetable or chicken stock,
   or water
2 tablespoons olive oil
4 garlic cloves, minced or pressed
2 medium onions, thinly sliced
2 medium tomatoes, sliced
2 red bell peppers, seeded and
   thinly sliced
2 green bell peppers, seeded and
   thinly sliced

2 cups chickpeas, cooked
1 teaspoon saffron threads
1 bay leaf
½ teaspoon salt
2 cups fresh green peas, lightly
   steamed until bright green,
   or 1 package frozen peas,
   thawed
6 ounces artichoke hearts,
   packed in water

Cook the rice in 3 cups of the stock until liquid is absorbed.

Heat the olive oil in a large skillet, flameproof casserole, or wok. Add the garlic, onions, tomatoes, and peppers and sauté until onions are translucent. Stir in the rice, remaining 3 cups of stock, chickpeas, saffron, bay leaf, and salt. Cover and simmer until water is almost absorbed, about 30 minutes.

Add green peas. Remove bay leaf. Garnish with artichoke hearts.

Make 6 servings.
279 calories per serving.

Delectable Butternut Squash

2 cups butternut squash, peeled
   and coarsely diced
1 large onion, chopped
½ cup carrot, thinly sliced
1 cup water

1½ cups kale, stemmed and
   coarsely chopped
3 tablespoons white miso*
⅓ cup fresh orange juice (about
   1 orange)

Simmer the vegetables in 1 cup of water in a medium saucepan for 4 minutes. Add the kale; cook for about 10 more minutes.

Dissolve the miso in the orange juice. Add the miso mixture to vegetables. Simmer for 2 minutes.

Makes 6 servings.
52 calories per serving.

*Miso can be purchased at health-food stores.

Sesame Cabbage

2 tablespoons sesame oil
2 tablespoons sesame seeds
2 pounds bok choy (Chinese
    cabbage) or Savoy cabbage,
    thinly sliced

¼ teaspoon salt

Heat the sesame oil in a wok or large nonstick saucepan over medium heat; add sesame seeds. Toast the sesame seeds for 2 minutes, stirring occasionally. Add the cabbage and salt. Cook until the cabbage wilts, about 5 minutes, stirring frequently.

Makes 4 servings.
112 calories per serving.

String Beans with Herbs

1 pound string beans
2 tablespoons olive oil
4 shallots, minced
1 garlic clove, minced or pressed
    (optional)
3 tablespoons fresh
    basil leaves, minced

½ teaspoon dried oregano
2 tablespoons fresh lemon
    juice
3 tablespoons fresh parsley,
    chopped

Cut the beans diagonally into 1½-inch pieces. Put them into a large saucepan and cover with cold water. Bring to a boil and cook until the beans are tender, about 6 to 8 minutes. Drain and rinse under cold water. Set aside.

Heat the olive oil in a medium skillet. Add the shallots and garlic, and sauté until the shallots are translucent. Stir in the basil and oregano and cook for 30 seconds. Add the beans, turning to make sure they become coated with the oil and herbs.

Add the lemon juice and parsley.

Makes 4 servings.
96 calories per serving.

Swiss Chard

1 pound Swiss chard	½ jalapeño pepper, cored, seeded,
2 tablespoons vegetable oil	and sliced
1 garlic clove, minced or pressed	½ cup chicken or vegetable stock
¼ cup shallots, minced	1 teaspoon salt

Bring a large pot of water to a boil. Wash the chard well. Cut the green leaves from the white ribs.

Drop the green leaves into the boiling water and cook for 4 minutes. Remove the leaves from the water (save the water for the white ribs); rinse the leaves under cool water. Drain, pat dry, and coarsely chop.

Cut the white ribs into 3-inch pieces. Drop them into the reserved boiling water. Cook for 15 minutes. Rinse under cold water. Drain and pat dry.

Heat the vegetable oil in a medium stockpot. Add the garlic, shallots, and jalapeño pepper; sauté until lightly browned, about 2 minutes. Add the greens and sauté for an additional 1 minute. Add the stock and salt, and cook 5 more minutes. Add the white stems and cook 5 more minutes.

Makes 4 servings.
90 calories per serving.

POULTRY

Lemon Chicken

3 teaspoons unsalted butter
1 pound boneless, skinless
   chicken breast, cut into
   bite-size pieces

½ cup dry white wine
juice of 2 lemons
2 lemons, sliced very thin
salt and pepper to taste

Melt the butter in a medium nonstick frying pan. Add the chicken pieces and sauté for 2 to 3 minutes, turning each piece once. Season with the salt and pepper, then add the wine, lemon juice, and lemon slices. Cover and simmer for 5 minutes.

Garnish with lemon slices.

Makes 4 servings.
165 calories per serving.

Savory Roast Chicken

4 skinless chicken breasts
¾ cup water
¼ cup soy sauce or tamari

1½ teaspoons garlic powder
1½ teaspoons dried rosemary
1 bay leaf

Preheat oven to 375°. Place the chicken in a roasting pan. Pour water and soy sauce over chicken, then add garlic powder and rosemary. Place bay leaf in the water.

Bake for 1½ hours, basting frequently until chicken is done. Add more water and soy sauce if necessary. Discard bay leaf.

Makes 4 servings.
304 calories per serving.

Baked Chicken with Vegetables

4 skinless chicken breasts
1 to 2 garlic cloves, crushed
½ to 1 teaspoon dried oregano
Salt and pepper to taste
2 medium zucchinis or summer
 squash, sliced

1 red bell pepper, sliced into
 strips
½ cup chicken stock
¼ cup white wine
4 tablespoons lemon juice

Preheat oven to 350°.

Place the chicken in a casserole dish. Sprinkle the chicken with garlic, oregano, salt, and pepper; then layer the vegetables on top.

In a small bowl mix together the chicken stock, wine, and lemon juice. Pour mixture over chicken.

Cover casserole tightly and bake until chicken is cooked through, about 30 minutes.

Makes 4 servings.
327 calories per serving.

Chicken, Broccoli, and Black Bean Stirfry

2 tablespoons vegetable oil
1 shallot, minced
2 boneless chicken breasts, sliced
 into ½-inch strips
2 cups broccoli florets, cut in
 small pieces

½ cup black beans, cooked
1 tablespoon hoisin sauce*
1 tablespoon soy sauce or tamari
pinch cayenne pepper
black pepper to taste (optional)

Heat the oil in wok or nonstick skillet. Add the shallot, chicken strips, and broccoli and cook over high heat, stirring constantly. Cook until the chicken is tender; then add the rest of the ingredients. Cook for 1 to 2 minutes until flavors are blended.

Add extra soy sauce and, if desired, freshly ground pepper.

Makes 4 servings.
245 calories per serving.

*Hoisin sauce can be purchased at local Asian markets or health-food stores.

Moroccan Chicken and Vegetables

4 boneless, skinless chicken
  breast halves
salt to taste
2 tablespoons olive oil
1 medium onion, thinly sliced
½ carrot, thinly sliced

2 garlic cloves, minced or pressed
1 can (8 ounces) whole tomatoes, including juice
1 to 2 cups fresh or frozen peas
¼ teaspoon ground cinnamon
black pepper to taste

Sprinkle the chicken with salt. Heat the olive oil in a large nonstick skillet; add the chicken. Sauté until lightly browned on both sides, about 5 minutes. Remove the chicken to a platter. Do not discard drippings.

To the drippings in the skillet add the onion, carrot, and garlic; sauté for 5 minutes, stirring occasionally. Stir in tomatoes, peas, cinnamon, and pepper, breaking up tomatoes with a spoon. Bring to a boil. Adjust seasoning.

Return the chicken to the skillet, mixing with sauce. Reduce heat to low; cover and simmer until chicken is cooked through, about 10 minutes.

Makes about 4 servings.
284 calories per serving.

Chicken with Garlic Sauce

1 tablespoon olive oil
4 garlic cloves, minced or pressed
4 boneless, skinless chicken
  breast halves

½ cup dry white wine
¼ cup fresh Italian parsley,
  chopped

Heat the olive oil in a large nonstick frying pan. Add the garlic and sauté for 1 minute. Add the chicken and sauté for about 2 minutes on each side.

Add the wine and bring to a boil. Lower heat, cover the pan, and cook until the inside of the chicken looks done, about 5 minutes. Do not overcook.

Garnish with fresh parsley.

Makes 4 servings.
338 calories per serving.

FISH

Poached Seafood

Choose seafood for poaching:
scallops, shrimp, squid, or
    shucked oysters
whole fish as poacher size allows
fish fillets and steaks (avoid meaty
    fish such as tuna, swordfish,
    and shark)

Choose for flavoring:
seasonings of choice
wine
fish stock

Combine water (or other liquid) and seasonings in a broad, shallow pan, checking that the liquid is deep enough to cover the seafood.

Bring the liquid to a boil, then reduce the heat. Add the seafood.

Poach, uncovered, until the seafood is opaque through the thickest part (cut to test). Transfer the seafood to a plate; cover to keep warm and set aside.

Ladle some cooking liquid through a strainer into a small pan and boil to reduce by half. Season as desired. Spoon the sauce over the seafood and serve.

Approximately 250 calories for a fish portion of about 6 oz.

Poached Fish with White Wine and Fennel

2 bulbs fennel
2 tablespoons olive oil
2 medium onions, thinly sliced
1 garlic clove, minced or
    pressed
½ teaspoon fennel seeds

1 large can (about 28 oz.)
    diced tomatoes
1 cup dry white wine
1½ pounds of 1-inch-thick fillets
    (e.g., cod, scrod, rockfish)
salt and pepper to taste

Rinse the fennel well. Trim and discard bulb bases. Cut off coarse tops of stalks, reserving a few green leaves for garnish. Cut bulbs lengthwise into quarters, then thinly slice crosswise.

Heat the olive oil in a 6-quart pan; add the sliced fennel and onions, sauté until onions are golden, about 8 to 10 minutes. Add garlic, fennel seeds, tomatoes, and wine. Bring to a boil, then cook until sauce is reduced, about 25 minutes.

Meanwhile, rinse and cut fish into 4 slices.

Add fish to tomato sauce. Reduce heat, cover, and simmer until fish is opaque, about 10 minutes. Do not overcook.

Salt and pepper to taste, and garnish with green fennel leaves.

Makes 4 servings.
254 calories per serving.

Steamed Flounder with Herbs

1½ pounds fillet of fish
   (scrod, sole, flounder)
black pepper to taste
basil, fresh or dried, to taste
parsley, fresh or dried, to taste
1 to 2 garlic cloves, minced
3 tablespoons white wine

1 tablespoon fresh lemon juice
1 tomato, sliced
1 medium onion, thinly sliced
   into rings
1 red bell pepper, thinly sliced
   rings
1 green bell pepper, optional

Preheat oven to 350°. Oil the bottom of a shallow pan. Layer the fish. Season with the pepper, basil, parsley, garlic, white wine, and lemon juice. Layer tomato, onion, and peppers.

Cover tightly with aluminum foil.

Cook about 25 minutes. Test to see if the fish is flaky for doneness. Remove carefully onto serving platter.

Makes 4 servings.
178 calories per serving.

Shrimp and Pea Pod Stirfry

2 tablespoons soy sauce or tamari
1½ teaspoons cornstarch
¼ cup sherry
¼ cup rice vinegar or white
   wine vinegar
1 tablespoon fresh ginger,
   finely chopped
2 tablespoons vegetable or olive oil

1 cup mushrooms, sliced
2 garlic cloves, minced or pressed
1 pound medium-size shrimp,
   shelled and deveined
¾ pound snow peas, ends and
   strings removed
¼ cup scallions, thinly sliced

In a small bowl, stir together the soy sauce, cornstarch, sherry, vinegar, and ginger. Set aside.

Heat 1 tablespoon of the vegetable oil in a wok or large, nonstick frying pan over high heat. Add the mushrooms and cook, stirring, until lightly browned, about 4 minutes. Stir in the garlic. Remove mixture to a separate bowl and set aside.

Heat the remaining 1 tablespoon of oil in pan. Add the shrimp and cook, stirring, until opaque in center, about 3 minutes. Add the mushroom mixture to wok; add pea pods and soy sauce mixture. Cook, stirring constantly, until pea pods turn bright green, about 2 minutes.

Garnish with scallions.

Makes 4 servings.
410 calories per serving.

Shrimp with Spicy Black Bean Sauce

2 cups black beans, cooked
1 tablespoon sesame or vegetable
   oil
1 teaspoon garlic, minced
   or pressed (optional)
1 tablespoon fresh ginger, minced
½ teaspoon crushed red pepper
1 pound medium shrimp, peeled,
   deveined, with tails left on

½ cup chicken or vegetable
   broth
2 teaspoons cornstarch
2 tablespoons soy sauce or
   tamari
1 tablespoon rice wine vinegar
4 scallions, finely chopped

In a medium mixing bowl, mash the beans with a fork.

Heat the oil in a wok or large skillet over medium-high heat. Add the garlic, ginger, and red pepper. Stirfry for 1 minute. Stir in the shrimp, beans, and chicken broth. Stir and cook until shrimp are opaque, about 5 minutes. Do not overcook.

Meanwhile, as the shrimp are cooking, combine cornstarch, soy sauce, and vinegar in a small bowl until well blended. Stir mixture into wok. Cook and stir until sauce boils and thickens, about 3 minutes. Stir in scallions, and cook for 1 minute.

Makes 4 servings.
297 calories per serving.

Shrimp Fried Rice

2 tablespoons vegetable oil
2 zucchinis, sliced*
1 cup celery, chopped*
4 scallions, chopped
1 cup small cooked shrimp
5 cups cooked rice

1 egg, or 2 egg whites, lightly beaten
2 tablespoons soy sauce or tamari
pinch cayenne pepper (optional)

Heat the vegetable oil in a wok or nonstick skillet. Add the vegetables and scallions and cook over high heat until vegetables are crisp-tender. Add the rest of the ingredients and mix well until eggs are set, about 2 to 3 minutes.

*For even cooking, cut the vegetables into uniformly sized pieces. Smaller pieces will cook faster.

Makes 4 servings.
363 calories per serving.

Fish Vindaloo

3 to 5 dried red chili peppers, seeded
½ cup plus 1 tablespoon white
   wine vinegar
3 to 4 garlic cloves, minced
½ teaspoon cumin seeds

Four 8-ounce flounder or sole
   fillets
2 tablespoons olive oil
1 medium onion, finely chopped
salt to taste

Soak the red chilies in 1 tablespoon of the vinegar for 10 minutes. Drain.

Combine the chilies, garlic, and cumin seeds in a food processor to form a paste. Add the remaining vinegar to form a thin paste.

Rub the chili paste on both sides of the fish. Cover and let marinate for 20 minutes.

Meanwhile, heat the olive oil in a large, nonstick skillet; add onion and sauté for 2 minutes. Reduce the heat to low. Add a little water and cook until onion is golden brown, about 10 to 15 minutes. Add the flounder, turning to cook on both sides, until the fish almost flakes, about 2 to 3 minutes on each side. Salt to taste.

Makes 4 servings.
275 calories per serving.

Thai Scallops with Basil

1½ pounds large fresh scallops
4 tablespoons red curry paste*
2 tablespoons vegetable oil
2 cloves garlic, minced or pressed

2 tablespoons fish sauce*
½ teaspoon sugar
¼ cup fresh basil leaves,
   slivered

Combine the scallops and red curry paste and marinate for 15 minutes.

Heat the vegetable oil in a wok or large nonstick frying pan. Add the marinated scallops and stirfry for 1 minute. Add the fish sauce and sugar; stir and cook another minute. Do not overcook. Add more fish sauce if needed.

Stir in basil and remove from heat.

Makes 4 servings.
231 calories per serving.

*These items can be purchased at a specialty or health-food store.

Chinese Baked Fish

1 tablespoon soy sauce or tamari
2 teaspoons sesame oil
4 flounder, perch, or catfish fillets
   (about 4 ounces each)

1 scallion, thinly sliced
½ cup red bell pepper, diced
½ cup seeded cucumber, diced
1 tablespoon fresh ginger, minced

Preheat oven to 450°. In a small bowl, mix soy sauce and sesame oil until well combined. Set aside.

Lay fillets onto a well-oiled baking pan. Sprinkle each fillet evenly with some scallion, red bell pepper, cucumber, and ginger; then drizzle with 1 teaspoon of soy sauce mixture. Wrap pan tightly with aluminum foil.

Bake for 10 minutes.

Makes 4 servings.
160 calories per serving.

Baked Yogurt Fish Fillets

1½ cups plain nonfat yogurt
½ red bell pepper, finely chopped
3 scallions, finely chopped
1½ tablespoons lemon juice
2 tablespoons fresh parsley, chopped

2 tablespoons chives, chopped
1½ pounds firm white fish fillets,
   such as cod, haddock, scrod, or
   striped bass, cut into bite-sized
   pieces

Preheat oven to 350°. Combine the yogurt, red bell pepper, scallions, lemon juice, parsley, and chives in a small bowl. Blend well.

Lay fish fillets in well-oiled baking pan. Pour yogurt mixture on top of fish. Cook until flesh of fish turns opaque or solid in color, about 8 to 12 minutes, depending on thickness of fish.

Makes 4 servings.
189 calories per serving.

DESSERTS

Applesauce

6 apples
1 cup unsweetened apple juice

1 to 2 teaspoons cinnamon

Chop the apples into 1-inch cubes, discarding the cores. In a medium saucepan bring the apple juice to a boil. Add the apples and cinnamon, then reduce heat and simmer for about 20 minutes. Serve hot or cold.

Makes 4 servings.
122 calories per serving.

Apple-Rice Pudding

Due to a higher protein content, this should not be eaten for breakfast.

1⅓ cups low-fat milk
3 egg whites
2 tablespoons sugar
2 teaspoons vanilla
½ teaspoon orange peel, finely grated
½ teaspoon cinnamon

2 cups cooked rice, brown or white
2 cups cooking apples, peeled and
    chopped (approximately 2 to 3)
⅓ cup dried apricots,
    chopped (optional)
⅛ teaspoon nutmeg

Preheat oven to 325°. Combine milk, egg whites, sugar, vanilla, orange peel, and cinnamon. Beat with an electric mixer until smooth. Stir in rice, apples, and apricots. Sprinkle with nutmeg.

Spoon into 8 custard cups; then put cups in pan filled with hot water to 1-inch depth. Bake until browned, about 50 to 60 minutes.

Remove cups from water. Allow to cool.

Makes 8 servings.
119 calories per serving.

CONDIMENTS

Apple Butter

2 cups water or unsweetened   pinch salt
   apple cider vinegar
3 pounds apples, washed, peeled,
   cored, and sliced

Put the water or vinegar, apples, and salt in a pot and bring to a boil. Cover, reduce heat to medium, and simmer until apples are soft. Remove apples from water and puree in a food processor, blender, or hand mill.

Place puree back in pot and simmer on low heat until apples become thick and dark, and the water has evaporated.

Placed in a jar, it will keep about 2 weeks in the refrigerator. Use it in place of jam or as a sweetener in rice, oatmeal, or barley.

Makes about 8 ounces.
25 calories per tablespoon.

BREADS

Almost all commercial breads are Bad Calories that attack a woman's serotonin. Although bread baking takes time it is well worth the effort, and when using a bread machine it's as simple as measuring ingredients. However, each bread machine requires that the ingredients be added in a specific order, that is, either the dry or the wet ingredients first. Also, some machines require more or less yeast than an average recipe. Adjust these recipes accordingly.

For the adventuresome who want to create new recipes themselves, here are some important guidelines:

1. A Good Calorie bread is one in which a predominance of the ingredients is made up of Good Calories. *At least* 50 percent of the dry ingredients should be whole-kernel grains and fiber such as: barley, cracked wheat (bulgur), wheat berries, rye berries, oats, oat bran, rice bran, wheat bran, and so on. Use other Good Calories (such as dried apricots).
2. Use vegetable oils instead of butter.
3. Guar gum is added to the breads and muffins because it has been proven to help lower the overall glycemic index when added to carbohydrates. It is important to use a prehydrated guar so that the recipe does not become gummy. See Appendix III, Mail-Order Resources.
4. Egg Replacer (by Ener-G) is an egg substitute. We use it in certain recipes that would be used for breakfast—for example, muffins—to keep the protein content low. If the foods from these recipes were to be eaten later in the day, regular eggs could be substituted. Egg Replacer can be purchased in your local health-food or specialty store.
5. Because these breads are denser, they work better if they have been kneaded a little longer than the normal bread-machine kneading cycle. To accomplish this, simply turn your machine off after ten minutes and start the whole cycle from the begin-

ning. This will give the dough an additional ten minutes of kneading.

6. For a lighter bread, remove dough after the end of the bread machine's two rising cycles and bake in 350° oven for 35 to 40 minutes.

We recommend eating only two servings of bread/muffins per day whether on the free-eating or caloric-restriction program.

Good Calorie Cracked Wheat Bread

Cracked wheat is the unprocessed wheat berry, or kernel, that has been crushed into pieces. It is also called bulgur.

1¼ cups water
¾ cup cracked wheat (bulgur)
¾ cup oats
1½ cups whole wheat flour
2 tablespoons gluten flour
1½ teaspoons salt

2 tablespoons olive oil
1 tablespoon molasses
2 teaspoons brown sugar
1½ teaspoons yeast
4 teaspoons guar gum

Bring the water to a boil; add the cracked wheat and oats. Reduce heat and simmer for 10 minutes. Remove from heat and let stand for about half an hour.

Then add the ingredients according to your individual bread machine's instructions. Set on whole wheat cycle.

Alternative method: For a lighter bread, remove dough after the end of the bread machine's two rising cycles and bake in a 350° oven for 35 to 40 minutes.

Remove from pan, place on rack, and allow to cool before slicing.

Makes 1 large loaf.
115 calories per slice.

Conventional method:

Bring 1 cup of the water to a boil; add the cracked wheat and oats. Reduce heat and simmer for 10 minutes. Remove from heat and let stand for about half an hour.

Stir in salt, olive oil, molasses, and brown sugar. Dissolve the yeast for 3 minutes in the remaining ¼ cup of warm water. Combine the cooked grains and yeast mixture. Then gradually beat in the whole wheat flour, gluten, and guar gum.

Turn the dough onto a floured surface and knead for about 10 minutes. Place dough into a greased bowl, cover, and let rise in a warm place until dough doubles in size, about 1 to 1½ hours. Punch dough in the center and work the edges of the dough into the center and turn the bottom of the dough to the top. Then place dough into greased 9-inch bread pan and let it rise, covered, until dough almost doubles in size again, about 1 to 1½ hours.

Preheat oven to 350°. Bake about 35 to 40 minutes.

(If you are using the conventional method, it works best to double the recipe and make two loaves.)

Good Calorie Cracked Wheat Focaccio

This is a variation on the Good Calorie Cracked Wheat Bread.

1¼ cups water
¾ cup cracked wheat (bulgur)
¾ cup oats
1½ cups whole wheat flour
2 tablespoons gluten flour
1½ teaspoons salt
2 tablespoons olive oil
1 tablespoon molasses
2 teaspoons brown sugar
1½ teaspoons yeast
4 teaspoons guar gum

Topping:
olive oil, enough to brush on
long-cooking oats, enough to
    sprinkle on top of bread
cornmeal, enough to sprinkle
    on top of bread
sunflower seeds, enough to
    sprinkle on top of bread

The bread machine is used only for mixing and rising in this recipe.

Add the ingredients according to the instructions given in your bread machine. Remove the dough from the machine after it's risen.

Preheat oven to 400°. Using your hands, flatten the dough on a surface sprinkled with oats. Flatten the dough so that it is about 2 inches thick. Coat the top of the bread with olive oil, using either a brush or your hands. Sprinkle the top with oats, cornmeal, and sun-flower seeds. Bake until top is golden brown, about 20 minutes.

Makes 1 loaf.
115 calories per slice.

The following four bread recipes were created by Dotti Lipetz. She is a fine artist by profession, not a baker, but she graciously offered to apply her considerable creative talents to bread making. We're thrilled she did. The results are delicious.

Good Calorie Sauerkraut-Rye Bread

1¼ cup oats
¾ cup water
1 cup sauerkraut, drained and cut
    in small pieces
¾ cup rye flour
¾ cup white flour
¼ cup oat bran
2 tablespoons gluten flour

1½ teaspoons salt
2 tablespoons olive oil
1½ tablespoons molasses
1½ tablespoons brown sugar
1 tablespoon caraway seeds
4 teaspoons guar gum
1½ teaspoons yeast

Soak the oats in the water for 20 to 25 minutes.

Then add the ingredients according to your individual bread machine's instructions. Set on whole wheat cycle.

Remove from pan, place on rack, and allow to cool before slicing.

Makes 1 large loaf.
115 calories per slice.

Good Calorie Oat Bread

1½ cup oats
1¼ cups warm water
½ cup oat bran
¾ cup bread flour
¾ cup oat flour
3 tablespoons gluten

1 tablespoon brown sugar
2 tablespoons vegetable oil
4 teaspoons guar gum
¾ teaspoon salt
1½ teaspoons yeast

Soak the oats in the warm water for 20 minutes.

Then add the ingredients according to your individual bread machine's instructions. Set on medium setting.

Remove from pan, place on rack, and allow to cool before slicing.

Makes 1 small loaf.
132 calories per slice.

Good Calorie Apricot-Orange Bread

¾ cup warmed orange juice
1¼ cups plus 1 tablespoon oats
¼ cup oat bran
½ cup bread flour
2 tablespoons gluten flour
1 tablespoon brown sugar

1 tablespoon vegetable oil
4 teaspoons guar gum
¾ teaspoon salt
2½ teaspoons yeast
½ cup dried apricots, slivered

Soak the oats in the warm orange juice for 20 minutes.

Then add the ingredients according to your individual bread machine's instructions. Set on medium cycle. Add apricots toward the end of the cycle.

Remove from pan, place on rack, and allow to cool before slicing.

Makes 1 small loaf.
110 calories per slice.

Good Calorie Applesauce Bread

1 cup warm water
2¼ cups oats
¾ cup oat bran
¼ cup bread flour
1 tablespoon gluten flour
1 tablespoon brown sugar

1 tablespoon vegetable oil
4 teaspoons guar gum
1 teaspoon salt
¾ teaspoon cinnamon
1½ teaspoons yeast
½ cup unsweetened applesauce

Soak the oats in the warm water for 20 minutes.

Then add the ingredients according to your individual bread machine's instructions. Set on medium cycle.

Remove from pan, place on rack, and allow to cool before slicing.

Makes 1 small loaf.
110 calories per slice.

MUFFINS

Good Calorie Apricot–Oat Bran Muffins

2¼ cups oat bran
5 teaspoons guar gum
1 tablespoon baking powder
½ teaspoon salt
2 Egg Replacer eggs
  (see page 238)

¾ cup orange, apple,
  or prune juice
⅓ cup molasses
¼ cup dried apricots,
  cut in slivers (optional)
2 tablespoons vegetable oil

Preheat oven to 425°. Mix oat bran, guar gum, baking powder, and salt. In a separate bowl mix Egg Replacer eggs, juice, molasses, apricots, and oil. Mix the dry ingredients into the liquid.

Line 12 muffin cups with cupcake liners. (These muffins work better with the liners.) Fill muffin tins three-quarters full with the batter. Bake until golden brown, about 15 minutes.

Makes 12 muffins.
110 calories per muffin.

Good Calorie Bran Muffins

1 cup whole wheat or white flour
1½ cups wheat bran
5 teaspoons guar gum
1 teaspoon baking soda
½ teaspoon salt
1 Egg Replacer egg (see page 238)

1 cup apple juice
2 tablespoons molasses
2 tablespoons brown sugar
2 tablespoons vegetable oil
1 cup fresh or frozen blueberries
  (optional)

Preheat oven to 375°. Combine the flour, bran, guar gum, baking soda, and salt. Mix. In a separate bowl, combine Egg Replacer egg, apple juice, molasses, brown sugar, and oil. Mix the dry ingredients into the liquid. Fold in blueberries.

Fill 12 greased muffin cups three-quarters full with the batter.

Bake until a wooden toothpick inserted comes clean, about 18 to 20 minutes. If using blueberries, bake for an additional 3 to 5 minutes.

Makes 12 muffins.
109 calories per muffin.

Good Calorie Oat Muffins

1¼ cups oat bran	¼ teaspoon salt
1 cup oats	3 teaspoons guar gum
1¼ cups all-purpose flour	2 Egg Replacer eggs
½ cup brown sugar	(see page 238)
1 teaspoon baking powder	1½ cups unsweetened applesauce

Preheat oven to 375°. Combine the oat bran, oats, flour, sugar, baking powder, salt, and guar gum. Mix. Add Egg Replacer eggs and applesauce. Mix.

Fill 15 greased muffin cups three-quarters full with the batter. Bake until a wooden toothpick inserted comes clean, about 15 to 20 minutes.

Makes about 15 muffins.
107 calories per muffin.

PANCAKES, CAKES, AND COOKIES

Almost all baked goods are Bad Calories that attack a woman's serotonin. These recipes are acceptable, however.

Sweet Potato Pancakes

1 Egg Replacer egg (see page 238)
1 medium sweet potato,
   peeled and cubed

2 tablespoons oat bran
¼ teaspoon baking powder
2 teaspoons butter

Place Egg Replacer in blender and turn on low speed. Gradually add the sweet potato. Continue blending, adding oat bran and baking powder. When mixture is evenly blended and thick, spoon out onto a griddle. (If using a skillet or crêpe pan, you may have to grease it lightly between batches.)

When bubbles appear on upper surface, lift the cakes to be sure they are browned. Then turn cakes to the other side and cook until nicely browned, about half the total time. Top with butter.

Makes 4 pancakes.
50 calories per pancake.

Good Calorie Sponge Cake

As this cake is high in protein, do not eat it for breakfast.

1 teaspoon lemon or
   orange rind, grated
¾ cup sugar
4 egg yolks
¼ cup boiling water
1 cup cake flour

1½ teaspoons baking powder
3 teaspoons guar gum
¼ teaspoon salt
1 teaspoon vanilla
4 egg whites

245

Preheat oven to 350°.

In a small bowl, stir the lemon or orange rind into the sugar.

With an electric or rotary beater, beat the egg yolks until very light. Beat in sugar gradually. Then beat in boiling water. Let cool.

Meanwhile, sift the flour. Then sift the baking powder, guar gum, and salt into the flour.

Once the egg yolks are cool, beat in the vanilla. Gradually add the sifted ingredients. Stir until blended.

Beat the egg whites until stiff. Fold them into the batter.

Pour the batter into a 9-inch tube pan. Do not grease pan; if there is any grease on the pan the batter will not rise.

Bake for 45 minutes.

Makes 8 slices.
159 calories per slice.

Good Calorie Oatmeal Cookies

1½ cups oats
½ cup whole wheat or white flour
⅜ cup sugar
4 teaspoons guar gum
1 teaspoon baking soda

½ cup unsweetened applesauce
1 beaten egg
1 teaspoon vanilla extract
½ teaspoon ground nutmeg

Preheat oven to 350°.

Combine the oats, flour, sugar, guar gum, and baking soda. Mix. In a separate bowl, combine the applesauce, egg, vanilla extract, and nutmeg. Mix well. Combine the two mixtures until the dough sticks together.

Drop rounded teaspoons of dough onto a greased cookie sheet. Slightly flatten each cookie with a fork.

Bake until golden brown, about 18 to 20 minutes.

Makes about 19 cookies.
65 calories per cookie.

Notes

CHAPTER 1 NOTES

1. Carlsson, M. et al. *J. Neural Transmission* **63:** 297–313 (1985).
   Giulian, D. et al. *Endocrinology* **93:** 1329–35 (1973).
   Kueng, W. et al. *Neuroendocrinology* **21:** 289–96 (1976).
   Carlsson, M. et al. *J. Neural Transmission* **63:** 297–313 (1985).
   Asberg, M. et al. *Clin. Pharmacology Therapeutics* **14:** 277–86 (1972).
2. Carlsson, M. et al. *J. Neural Transmission* **63:** 297–313 (1985).
3. Giulian, D. et al. *Endocrinology* **93:** 1329–35 (1973).
   Kueng, W. et al. *Neuroendocrinology* **21:** 289–96 (1976).
   Carlsson, M. et al. *J. Neural Transmission* **63:** 297–313 (1985).
4. Carlsson, M. et al. *J. Neural Transmission* **63:** 297–313 (1985).
5. Amit, Z. et al. **52 (suppl 12):** 55–60 (1991).
6. Young, S. N. et al. *J. of Neurology, Neurosurgrey, and Psychiatry* **43:** 438–45 (1980).
   Vaccari, A. et al. *Brain Res.* **132:** 176–85 (1977).
   Simon, N. & Volicer, Z. L. *J. Neurochem.* **26:** 893–900 (1976).
   Skillen, R. G. et al. *Endocrinology* **69:** 1099–1102 (1961).
   Hardin, C. M. *Brain Res.* **59:** 437–39 (1973).
7. Dickinson, S. L. & Curzon, G. *Neuropharmaocology* **25:** 771–76 (1986).
8. Rodin, J. et al. *Metabolism* **34:** 826–31 (1985).
9. Slabber, M. et al. *Am. J. Clin. Nutr.* **60:** 48–53 (1994).

CHAPTER 2 NOTES

1. Walsh, A. E. S. et al. *J. Affective Disorders* **33:** 89–97 (1995).
2. Anderson, I. M. et al. *Psychological Med.* **20:** 785–91 (1990).
   Goodwin, G. M. et al. *Arch. Gen. Psychiatry* **44:** 952–57 (1990).
3. Delgado, P. L. et al. *Life Sciences* **45:** 2323–32 (1989).
   Schweiger, U. et al. *J. Neural Transmision* **67:** 77–86 (1986).
   del Barrio, A. S. et al. *Nutr. Res.* **16:** 167–78 (1996).
   McCargar, L. J. et al. *Am. J. Clin. Nutr.* **47:** 932–34 (1988).
4. Kaptein, E. M. et al. *Clin. Endocrinol.* **22:** 1–15 (1985).
   Rozen, R. et al. *Int. J. Obesity.* **10:** 303–12 (1986).
5. Schweiger, U. et al. *J. Neural Transmission* **67:** 77–86 (1986).
   Maes, M. et al. *J. Affective Disorders* **15:** 119–25 (1988).
6. Goodwin, G. M. et al. *J. Affective Disorders* **14:** 137–44 (1988).
   Loosen, P. T. & Prange, A. *J. Am. J. Psychiatry* **139:** 405–16 (1982).
   Kirkegard, C. *Psychoneuroendocrinology* **6:** 189–212 (1981).
   Kaptein, E. M. et al. *Clin. Endocrinol.* **22:** 1–15 (1985).
   Rozen, R. et al. *Int. J. Obesity.* **10:** 303–12 (1986).
7. Loosen, P. T. & Prange, A. *J. Am. J. Psychiatry* **139:** 405–16 (1982).
   Kirkegard, C. *Psychoneuroendocrinology* **6:** 189–212 (1981).
   Goodwin, G. M. et al. *J. Affective Disorders* **14:** 137–44 (1988).
8. Carroll, B. J. et al. *Arch. Gen. Psychiatry* **38:** 15–22 (1981).
   Berger, M. et al. *Arch. Gen. Psychiatry* **40:** 585–86 (1983).
   Mullen, P. E. et al. *Lancet ii*: 1051–55 (1986).
   Fitchter, M. M. et al. *Psychiatry Res.* **17:** 61–72 (1986).
   Goodwin, G. M. et al. *J. Affective Disorders* **14:** 137–44 (1988).
9. Walsh, A. E. S. et al. *J. Affective Disorders* **33:** 89–97 (1995).
10. Keys, A. et al. *The Biology of Human Starvation* 2 (1952).
11. Green, M. W. et al. *Physiology & Behavior* **55:** 447–52 (1994).
12. Green, M. W. et al. *Physiology & Behavior* **55:** 447–52 (1994).
13. Kwok, R. P. S. & Juorio, A. V. *Neuroendocrinology* **45:** 267–73
   (1987).
   Wurtman, R. J. & Wurtman, J. J. *Appetite* **7:** supplement, 99–103
   (1986).
   Ashley, D. V. M. et al. *Am. J. Clin. Nutr.* **42:** 1240–45 (1985).
   Paykfl, S. et al. *Br. J. Psychiat.* **123:** 501–7 (1973).
   Blum, I. et al. *Am. J. Clin. Nutr.* **57:** 486–89 (1993).
   Caballero, B., & Wurthman, R. J. *Metabolism* **40:** 51–58 (1991).
   Caballero, B. et al. *Metabolism* **37:** 672–76 (1988).

Caballero, B. et al. In: *Amino Acids in Health and Disease: New Perspectives* (New York: Alan R. Liss publishers, 1987), 369–82.

Ashley, D. V. M. et al. *Am. J. Clin. Nutr.* **42:** 1240–45 (1985).

Forlani, G. et al. *Metabolism* **33:** 147–50 (1984).

14. Hopkinson, G. & Bland, R. C. *Can. J. Psychiatry* **27:** 213–15 (1982).

15. Fernstrom, J. D. *Appetite* **8:** 163–82 (1987).

    Nutrition Reviews **45:** 87–89 (1987).

    Blum, I. et al. *Metabolism* **41:** 137–40 (1987).

    Fernstrom, J. D. & Wurtman, J. J. *Science* **178:** 414–16 (1972).

16. Rodin, J. et al. *Metabolism* **34:** 826–31 (1985).

17. Kowk, R. P. S. & Juorio, A. V. *Neuroendocrinology* **45:** 267–73 (1987).

18. Yamada, J. et al. *European J. Pharmacol.* **181:** 319–22 (1990).

19. Spring, B. et al. *Health Psychology* **10:** 216–23 (1991).

    Spring, B. et al. *Int. J. Obes. Relat. Metab. Disord.* 16 Suppl 3: S19–23 (1992).

    Bowen, D. J. et al. *Behavioral Med.* **14:** 97–110 (1991).

    Daoust, M. et al. *Alcohol & Alcoholism* **27:** 15–17 (1992).

    Levin, E. D. et al. *Pharmacol. Biochem. Behav.* **44:** 51–61 (1993).

    Pomerleau, O. F. et al. *Psychoneuroendocrinology* **16:** 433–40 (1991).

20. Krahn, D. et al. *J. Subt. Abuse* **4:** 341–53 (1992).

21. Sellers, E. M. et al. *J. Clin. Psychiatry* **52 (Suppl 12):** 49–54 (1991).

    Ahlikas, J. A. et al. *J. Addict. Dis.* **12:** 129–39 (1993).

    Maremmani, I. et al. *Br. J. Psychiatry* **160:** 57–71 (1992).

    Yu, D. S. et al. *Life Sci.* **13:** 1383–88 (1986).

22. Drewnowski, A. et al. *Physiol. Behav.* **51:** 371–79 (1992).

23. Blum, I. et al. *Metabolism* **41:** 137–40 (1987).

24. Muldoon, M. F. et al. *Biol. Psychiatry* **31:** 739–42 (1992).

25. Kolata, Gina, *New York Times*, April 25, 1995.

26. Anderson, I. M. et al. *Pyschological Med.* **20:** 785–91 (1990).

    Delgado, P. L. et al. *Life Sciences* **45:** 2323–32 (1989).

27. Anderson, I. M. et al. *Psychological Med.* **20:** 785–91 (1990).

    Goodwin, G. M. et al. *Arch. Gen. Psychiatry* **44:** 952–57 (1990).

CHAPTER 3 NOTES

1. Grunberg, N. E. et al. *Psychopharmacology* **83:** 93–98 (1984).

   Grunberg, N. E. et al. *Psychopharmacology* **87:** 198–203 (1985).

   Grunberg, N. E. et al. *Psychopharmacology* **90:** 101–05 (1986).

2. Scuteri, F. *Mongr. Neural Sci.* **3:** 94–101 (1976).

3. Brzenski, A. A. et al. *Obstet. Gynecol.* **76:** 296–301 (1990).
4. Sanday, P. R. *Female Power and Male Dominance: On the Origins of Sexual Inequality* (Cambridge, England: Cambridge University Press, 1981).
5. Leslie Laurence and Beth Weinhouse, *Outrageous Practices: The Alarming Truth About How Medicine Mistreats Women* (New York: Ballantine Books, 1994).

CHAPTER 4 NOTES

1. Amit, Z. et al. **52 (suppl 12):** 55–60 (1991).
2. Walsh, A. E. S. et al. *J. Affective Disorders* **33:** 89–97 (1995).
3. Miller, H. L. et al. *J. Clin. Psychiatry* **53 (suppl):** 28–35 (1992).
4. Walsh, A. E. S. et al. *J. Affective Disord.* **33:** 89–97 (1995).
5. del Barrio, A. S. et al. *Nutr. Res.* **16:** 1671–77 (1996).
6. Finer, N. et al. *Clinical Neuropharmacology* **11:** S178–S186 (1988).
7. Goodwin, G. M. et al. *J. Affective Disord.* **30:** 117–22 (1994).
8. Ashley, D. V. M. et al. *Am. J. Clin. Nutr.* **42:** 1240–45 (1985).
9. Cangiano, et al. *Am. J. Clin. Nutr.* **56:** 863–67 (1992).

CHAPTER 5 NOTES

1. Jenkins, D. J. A. et al. *Am. J. Gastroenterolgy* **84:** 732–39 (1989).
   Liberman, H. R. et al. *J. Neural Transmission* **65:** 211–17 (1986).
   Christensen, L. & Redig, C. *Behavioral Neuroscience* **2:** 346–53 (1993).
   Lyons, P. M. & Trustwell, A. S. *Am. J. Clin. Nutr.* **47:** 433–39 (1988).
   Wolever, T. M. S. et al. *Am. J. Clin. Nutr.* **48:** 1041–47 (1988).
2. Holt, S. et al. *Appetite* **18:** 129–41 (1992).
   Leathwood, P., & Pollet, P. *Appetite* **10:** 1–11 (1988).
3. Knutowski, M. et al. *International J. Obesity* **16 (suppl 4):** S63–S66 (1992).
   Gray, D. S. et al. *International J. Obesity* **16 (suppl 4):** S67–S72 (1992).
   Potter van Loon, B. J. et al. *International J. Obesity* **16 (suppl 4):** S55–S61 (1992).
   Arora, R. et al. *Eur. J. Clin. Invest.* **24:** 182–87 (1994).
   Stewart, G. O. et al. *Med. J. Aust.* **158:** 167–69 (1993).
   Scheen, A. J. et al. *Diabetes Care* **14:** 325–32 (1991).

Proietto, J. et al. *Diabetes Res. Clin. Pract.* **23:** 127–34 (1994).

Anderson, P. H. et al. *Acta. Endocrinol. (Copenh.)* **128:** 251–58 (1993).

4. Blundell, J. E. *Int. J. Obes. Relat. Metab. Disord.* **16 (suppl 3):** S51–S59 (1992).

    Pijl, H. et al. *Int. J. Obes.* **15:** 237–42 (1991).

    Lafreniere, F. et al. *Int. J. Obes. Relat. Metab. Disord.* **17:** 25–30 (1993).

5. Wurtman, J. et al. *Neuropsychopharmacology* **9:** 201–10 (1993).

    Pijl, H. et al. *Int. J. Obes. Relat. Metab. Disord.* **17:** 513–20 (1993).

    Blundell, J. E. & Lawton, C. L. *Metabolism* **44 (2 suppl 2):** 33–37 (1995).

6. *Nutr. Rev.* **52 (2 pt 1):** 65–68 (1994).

7. Knutowski, M. et al. *International J. Obesity* 16 **(suppl 4):** S63–S66 (1992).

8. Goldstein, D. J. et al. *Int. J. Obes. Relat. Metab. Disord.* **18:** 129–35 (1994).

9. Bjorntorp, P. *Metabolism* **44 (suppl. 2):** 38–41 (1995).

10. Slabber, M. et al. *Am. J. Clin. Nutr.* **60:** 48–53 (1994).

11. Lipetz, P. D. *The Good Calorie Diet* (New York: HarperCollins, 1994).

12. Gulliford, M. C. et al. *Am J. Clin. Nutr.* **50:** 773–77 (1989).

    Gatti, E. et al. *Eur. J. Clin. Nutr.* **46:** 161–66 (1991).

    Collier, G. et al. *Diabetologia* **26:** 50–54 (1984).

    Himsworth, H. P. *Clin. Sci.* **2:** 67–94 (1935).

    Sweeney, J. S. *Arch. Intern. Med.* **40:** 818–30 (1927).

13. Jones, P. J. H. et al. *Metabolism* **41:** 396–401 (1992).

14. Schweiger, U. et al. *J. Neural Transmission* **67:** 77–86 (1986).

    Fernstrom, J. D. *Am. J. Clin. Nutr.* **42:** 1072–82 (1985).

15. Asheley, D. V. M. *Internat. J. Vit. Nutr. Res.* **suppl. 29:** 27–40 (1986).

16. Schweiger, U. et al. *J. Neural Transmission* **67:** 77–86 (1986).

    Fernstrom, J. D. *Am. J. Clin. Nutr.* **42:** 1072–82 (1985).

17. de Burgos, A. et al. *Eur. J. J. Clin. Nutr.* **46:** 803–8 (1992).

    DeFronzo, R. A. *Metabolism* **37:** 105–8 (1988).

    Kersteer, J. et al. *Metabolism* **40:** 707–13 (1991).

18. Truman, R. W. & Doisy, R. J. in *Trace Elements Metabol. in Animals–2,* Hoekstra, J. W. (ed.) 678 (1974).

    Anders, R. A. *Clin. Physiol. Biochem.* **4:** 31–41 (1986).

19. Glaser, E. & Halpern, G. *Biochem. Z.* **207:** 377–83 (1929).

20. Glinsmann, W. H. & Mertz, W. *Metabolism* **15:** 510–15 (1966).

    Levine, R. A. et al. *Metabolism* **17:** 114–25 (1968).

    Hopkins, L. L. & Price, M. G. *Nutrition Congress* **2:** 40 (1968).

21. Mertz, W. *Physiol. Rev.* **49:** 163–239 (1969).

22. Anderson, R. A. & Kozlovsky, A. S. *Am. J. Clin. Nutr.* **41:** 1177–83 (1985).

23. Anderson, R. A. et al. *Am. J. Clin. Nutr.* **51:** 864–68 (1990).

    Mertz, W. *Physiol. Rev.* **49:** 163–239 (1969).

    Anderson, R. A. *Clin. Physiol. Biochem.* **4:** 31–41 (1986).

    Kozlovsky, A. S. et al. *Metabolism* **35:** 515–18 (1986).

24. Hopkins, L. L. *Am. J. Clin. Nutrition* **21:** 203–11 (1968).

    Mertz, W. in *Present Knowledge in Nutrition* (Washington, D.C. : The Nutrition Foundation, 1976), 365–72.

25. Schwartz, K. & Mertz, W. *Arch. Biochem. Biophys.* **72:** 515–18 (1957).

26. Urberg, M. & Zemel, M. B. *Metabolism* **36:** 896–99 (1987).

27. Urberg, M. et al. *The J. Family Practice* **27:** 603–6 (1988).

    Lefavi, R. et al. *FASEB J.* **5:** A1645 (1991).

28. Lefavi, R. G. *Inter. J. Sports Nutr.* **3:** 120–21 (1993).

29. Lefavi, R. G. et al. *Inter. J. Sports Nutr.* **2:** 111–22 (1992).

    Clarkson, P. M. *Inter. J. Sports Nutr.* **1:** 289–93 (1991).

    Moore, R. J. & Friedl, K. E. *Nat. Strength Cond. Assoc.* J. 47–51 (1992).

    Lefavi, R. G. *Inter. J. Sports Nutr.* **3:** 120–21 (1993).

    Hallmark, M. A. et al. *Med. Sci. in Sports & Exercise* **25:** S101 (1993).

    Clancey, S. et al. *Med. Sci. in Sports & Exercise* **25:** S194 (1993).

    Hasten, D. L. et al. (Conference Abstract) SE Regional Chapter, Am. College of Sports Med. (1991).

30. Fernandez–Pol, J. A. & Johnson, G. S. *Cancer Res.* **37:** 4276–79 (1977).

31. Fernandez–Pol, J. A. *Exp. Mol. Path.* **29:** 348–57 (1978).

32. Etzel, K. R. et al. *Nutr. Res.* **8:** 1391–1401 (1988).

33. Fernandez–Pol, J. A. *Biochem. Biophys. Res. Comm.* **78:** 136–43 (1977).

34. Seal, C. J. *Ann. Nutr. Metbol.* **32:** 186–91 (1988).

35. Banners, S. *Nature* **152:** 152 (1943).

    Altenburger, E. *Klinishe Wochenschrift* **15:** 1129–31 (1936).

    Banerjee, S. *J. Biol. Chem.* **168:** 207–11 (1947).

36. Schorah, C. J. et al. *Internat. J. Vit. Res.* **58:** 312–18 (1988).
    King, C. G. *J. Biol. Chem.* **116:** 489–92 (1936).
    Banerjee, S. *J. Biol. Chem.* **168:** 207–11 (1947).
    Banners, S. *Nature* **152:** 152 (1943).
37. Kersteer, J. et al. *Metabolism* **40:** 707–13 (1991).
    Geddik, O. & Akalin, S. *Diabetolgica* **29:** 142–45 (1986).
38. Paolisso, G. et al. *Am. J. Clin. Nutr.* **57:** 650–56 (1993).
39. Durlach, J. et al. *Magnesium* **2:** 192–224 (1983).
    Paolisso, G. et al. *Diabetes Care* **12:** 265–69 (1989).
40. Rabinowitz, D. and Zierler, K. L. *J. Clin. Invest.* **41:** 2173–81 (1962).
    DeFronzo, R. A. *Metabolism* **37:** 105–8 (1988).
41. Schweuger, U. et al. *Metabolism* **35:** 938–43 (1986).
    Heraief, E. et al. *J. Neural Transm.* **57:** 187–95 (1983).
42. Schweuger, U. et al. *Metabolism* **35:** 938–43 (1986).
    Heraief, E. et al. *J. Neural Transm.* **57:** 187–95 (1983).
43. Teff, K. L. et al. *Pharmacol. Biochem. Behav.* **34:** 829–37 (1989).
    Christensen, L. & Redig, C. *Behav. Neurosci.* **107:** 346–53 (1993).
    Moller, S. E. *J. Neural Transm.* **76:** 55–63 (1989).
    Pijl, H. et al. *Intern. J. Obesity* **17:** 513–20 (1993).
44. Luo, S. & Li, E. T. S. *Psychopharmacology* **109:** 212–16 (1992).
45. Williamson, D. J. et al. *British J. Psychiatry* **167:** 238–42 (1995).
46. Moses, P. L. & Wurtman, R. J. *Life Sci.* **35:** 1297–1300 (1984).
47. Williamson, D. J. et al. *British J. Psychiatry* **167:** 238–42 (1995).

CHAPTER 6 NOTES

1. O'Donell, L. J. D. et al. *BMJ* **298:** 1616–17 (1989).
2. Ellis, P. R. et al. *Br. J. Nutr.* **46:** 267–76 (1981).
3. Herxberg, G. R. & Rogerson, M. *J. Nutr.* **118:** 1061–67 (1988).
4. Higgins, H. L. *Am. J. Physiol.* **41:** 258–65 (1916).
5. Herxberg, G. R. & Rogerson, M. *J. Nutr.* **118:** 1061–67 (1988).
6. Hallfrisch, J. et al. *J. Nutr.* **113:** 1819–26 (1983).
   Reiser, S. et al. *Am. J. Clin. Nutr.* **45:** 580–87 (1987).
   Thorburn, A. W. et al. *Am J. Clin. Nutr.* **49:** 1155–63 (1989).

CHAPTER 9 NOTES

1. Chaouloff, F. et al. *Neurorendocrinology* **50:** 344–50 (1989).
2. Kennett, G. A. et al. *Brain Research* **382:** 416–21 (1986).

3. Heinsbroek, R. P. W. et al. *Pharmacol. Biochem. Behavior* **37:** 539–50 (1990).
4. Higley, J. D. et al. *Psychopharmacology* **103:** 551–56 (1991).
5. Heinsbroek, R. P. W. et al. *Pharmacol. Biochem. Behavior* **31:** 499–503 (1988).
6. Hagstrom-Toft, E. et al. *J. Clin. Endrocrin. Metabol.* **76:** 392–98 (1993).
7. Kahn, C. R. et al. *Endocrinology* **10:** 1054, 66 (1978).
8. Anderson, I. M. et al. *Psychological Med.* **20:** 785–91 (1990).
9. Chaouloff, F. *Acta Physiol. Scand.* **137:** 1–13 (1989).
   DeCoverly Veale, D. M. V. *Acata Physiol. Scand.* **76:** 113–20 (1987).
   Dey, S. *Physiology & Behavior* **55:** 323–29 (1994).
   Brown, J. D. & Lawton, M. *J. Human Stress* **12:** 125–31 (1986).
   Goodyear, L. J. et al. *Metabolism* **40:** 455–64 (1991).
   Tan, M. H. et al. *J. Appl. Physiol.* **52:** 1514–18 (1982).
   Byrene, A. & Byrene, D. G. *J. Psychosomatic Res.* **37:** 565–74 (1993).
10. Goodyear, L. J. et al. *Metabolism* **40:** 455–64 (1991).
    Tan, M. H. et al. *J. Appl. Physiol.* **52:** 1514–18 (1982).
11. Dey, S. *Physiology & Behavior* **55:** 323–29 (1994).
    Chaouloff, F. *Acta Physiol. Scand.* **137:** 1–13 (1989).
    Dey, S. et al. *Physiological Behavior* **52:** 1095– 99 (1992).
12. Dey, S. *Physiology & Behavior* **55:** 323–29 (1994).
13. Sudsuang, R. et al. *Physiology & Behavior* **50:** 543–48 (1991).
14. Werner, R. G. et al. *Psychosomatic Med.* **48:** 59–61 (1986).
    Sudsuang, R. et al. *Physiology & Behavior* **50:** 543–48 (1991).
15. Werner, R. G. et al. *Psychosomatic Med.* **48:** 59–61 (1986).
16. Bujatti, M. & Riederer, *J. Neural Transm.* **39:** 257–67 (1976).

Ana's Voice

When God Speaks
Through A Silent Child

by
Randy Jessen

authorHOUSE™

1663 LIBERTY DRIVE, SUITE 200
BLOOMINGTON, INDIANA 47403
(800) 839-8640
WWW.AUTHORHOUSE.COM

First published by AuthorHouse 12/16/05

ISBN: 1-4208-9169-3 (sc)
ISBN: 1-4208-9170-7 (dj)

Printed in the United States of America
Bloomington, Indiana

This book is printed on acid-free paper.

Dedication

This book is dedicated:

To Suzi and Ana

To the ministry of Global Hope and StarStone International
To all the children who find themselves lost and alone
To every family that struggles with HIV/AIDS
To every person who lives with a dream

To Flavius, Gabi, Teo, and Mircea

Acknowledgements

I offer my deepest thanks and absolute love to my extended family. Every person in that circle has been an instrumental part of this journey. I am especially grateful to my wife Sue. She has a wonderfully patient way of helping my projects find their way to completion. She is my partner in ministry and my soul mate for life. She is also the best mother and mentor a little girl could ever want.

In addition, my thanks goes out to our daughter and son in law, John and Cheri Walker. They played a key role in the unfolding drama of Ana's life. It was their personal compassion that initially opened the doors to the people and culture of Romania. It was also their passion and determination that saved Ana's life and ultimately brought her into our home. Sue and I are eternally grateful.

Our journey was strengthened by the support of our mothers, Louise Dozier and Liz Jessen. Our children David and Kim Brady remain valued helpers and ready assistants. Our grand children Tara, Davis, Kyler, Loredana, Miamina, Sam and Keyla taught Ana to play, laugh, and love.

I also know that the journey of any family is always played out in the context of extended relationships. For us, those relationships have taken shape through the church, school systems, medical professionals, a passionate therapeutic community, and a host of friends. I wish I could introduce you personally to every individual. For now, please forgive me for simply offering a brief bullet point list

of friends and organizations. Each one has been an important part
of our journey and a living extension of our faith.

- The people of the Genoa United Methodist Church who loved us
  to the Lord and brought light to our lives
- Pastor Al and Flo Unger and the people of the Fort Morgan
  United Methodist Church
- The faithful folks at Longs Peak United Methodist Church who
  started the journey with us
- Our faith community at the Broomfield United Methodist
  Church where Ana found her first home
- Our friends and family of faith at First United Methodist Church
  in Colorado Springs who adopted us into their lives
- Members of the CHIP Team at Children's Hospital in Denver
  who provided loving care for Ana and our family
- The faculty and graduate students at the University of Colorado
  Speech Language and Communication center who taught sign
  language, brought sound to life, and taught Ana how to play
- The teachers, staff and students at
  - Hulstrom Elementary School
  - Cotton Creek Elementary School
  - Westlake Middle School
  - Eagleview Middle School
  - Air Academy High School
- The people of the Speranta Baptist Church in Arad, Romania
  - Special thanks to Pastor Romica Iuga, Pastor Viorel Iuga
    and Brother Nelu Redis
- Our Bucharest friends coworkers
  - George and Mariana Cojocaru
  - Marius and Nicoleta Caramitru
  - Traian and Gabi Condrea
  - Dan and Rodica Nestor
- The children, Board of Directors, staff, and mission personnel of
  Global Hope, Inc.
- Team members who traveled with me to Romania to discover
  new ways to serve the Lord and returned with lives that were
  forever transformed
- The Board of Directors and chief dreamers of StarStone
  International
- To Rick Nelson who volunteered his time to edit the details and
  the folks at AuthorHouse for their support

Finally, my deepest thanks to the Lord Jesus Christ who opened the doors, provided encouragement, created the resources, and dried many tears. Mostly I thank the Lord for bringing Ana into our lives. Sue and I have been blessed by the life, love and laughter of this beautiful child of God. AnaBabe, thanks for giving me a chance to play on the floor. Thanks for teaching us how to laugh. Above all, thanks for sharing your life with us so we could have a wonderful first hand experience with the miracle of faith and the power of hope.

Contents

Dedication   ~   v
Acknowledgements   ~   vii

Chapter One
The Prelude   ~   1

Chapter Two
The Journey   ~   9

Chapter Three
The Breakthrough   ~   15

Chapter Four
Homeward Bound ~ Forever Changed   ~   39

Chapter Five
The Joy of Spiritual Discovery   ~   55

Chapter Six
Family Matters   ~   71

Chapter Seven
Silent Child   ~   79

Chapter Eight
Escape to Home   ~   87

Chapter Nine
New Life from the Breath of God   ~   97

After Thoughts   ~   105
Appendix   ~   109

Chapter One

The Prelude

For I know the plans I have for you, declares the Lord,
plans to prosper you and not to harm you,
plans to give you hope and a future.

Jeremiah 29:11 NIV

The year 1989 was filled with joy and turmoil. It was the year our daughter Ana was born. She came into a world of turmoil and pain on March 8, 1989. Her actual place of birth remains a mystery that is shrouded in two different accounts. Some say that her birth parents, who lived about 15 kilometers to the north, were visiting friends in the gypsy section of Arad, Romania known as Kekecs. Perhaps that tiny one-room house with a dirt floor and no water, in the city of Arad became her birthplace. Other accounts have her born in the Arad maternity hospital and removed from her birth family by government officials because the family was unable to care for her.

Either way, Ana Sue Jessen started her life as the ninth child born to a poor gypsy family in western Romania. Ana and Ovidiu Cionca brought her into a world that was about to turn upside down.

Nicolae Ceausescu had been the communist dictator of Romania since 1965. By the mid-1980s, the economy had been drained by Ceausescu's demented desire to build grand monuments to identify himself as a world-class leader. While people were standing in line

1

for hours to feed their families, he was selling off the resources of the country to build a huge palace in downtown Bucharest. His Bucharest palace was still under construction at the time of his death. While people were starving, he had nearly 700 architects designing his 900-room palace on Unirii Boulevard. The ostentatious splendor of Ceausescu's demented design was only rumored until the people discovered his excesses after the revolution of December 1989. The "Palace of Parliament," as it is now known, is the second-largest building in the world. It is second only to the Pentagon in Washington, DC.

Ceausescu destroyed the spirit of the people through his waste of national resources and his single-minded destruction of the economy to satisfy his need for recognition. He infiltrated the entire social structure with his private secret police and controlled every move of every person, from the aristocrats to the peasant farmers.

For years, Ceausescu had been one of the most paranoid Communist leaders. He was frightened that someone would poison him or try to overthrow him. The former head of Romanian Intelligence reported that Ceausescu had warehouses of new clothing placed all around the country. His fear of being poisoned forced him to demand a completely new set of clothing for each new day. Every pair of shoes, every suit, and every pair of socks were marked for destruction after he wore them once. Even so, in 1989 he seemed oblivious to the political upheaval that surrounded his country.

The 12-foot tall, 103-mile-long Berlin wall started to come down on November 9, 1989. It stood for 28 years as a symbol of oppression and the division between east and west. At the same time, the former Soviet Union was being crushed under the weight of a failed communist economy. One communist leader after another began to fall. But Ceausescu's mind was focused on meeting his own lavish needs. He was not able to learn from the political upheaval that surrounded him. It proved to be a fatal error. 1989 will forever be known as the year Communism crashed.

By the time Ana was born in March of that year, the mood of the Romanian people was turning from frustration to rebellion. Political and spiritual unrest first broke out in the streets of Timisoara. By December 21, 1989, there was open revolution in the streets of

Bucharest. On Christmas Day 1989, Nicolae Ceausescu and his wife Elena were captured, convicted, and executed. Twenty-four years of Communist oppression came to an end on that day.

But it wasn't over yet. The Romanian economy had been devastated.The people were starving. They were hungry for food, trust, and hope. There was a severe shortage of food. The Christian church had been oppressed for years, and the future held very little promise. Within days, the currency of the land had little value, the people had little hope for the future, and no one knew if their neighbor could be trusted.

This is the world into which our daughter was born. It was an environment of radical change, poverty, and pain. She couldn't have been more than a few days old when her father wrapped her in a blanket and started the 22-kilometer trip south from their home in Curtici to the county seat City of Arad. He took her directly to the state-operated orphanage on Vicentiu Babes Street—where he abandoned her. Mr. and Mrs. Cionca would next see their daughter when she was 5½ years old and on her way to America.

Some 250 other infants and very young toddlers lived at the Vicentiu Babes orphanage when Ana arrived. She was given a quick medical exam and found to be a bit under weight. She needed a little boost, so she was given an injection of about 3 milliliters of raw blood. They thought it would "pink her up" and give her a chance to catch up with the rest of the kids. There was only one problem—that small amount of blood was contaminated. She was infected with typhoid, a strain of hepatitis and the virus that causes AIDS (Acquired Immune Deficiency Syndrome). This misguided effort to help would change Ana's life forever. All this, and it was still 1989.

The Christmas revolution and execution of Nicolae and Elena Ceausescu barely made the evening news in the USA. The pages of Time and Newsweek carried small articles about fighting in the streets of Bucharest. Americans were captivated by the break-up of the Soviet Union. The collapse of the "evil empire" and the larger collapse of Communism took center stage. No one had time to pay much attention to another Eastern Block dictator biting the dust.

It didn't take long for the world to learn the real legacy of Ceausescu's distorted view of leadership. His paranoia and his quest for absolute power were powerfully demonstrated by thousands of little white iron cribs that were filled with abandoned children from one end of the country to the other.

Medical missionaries from Western Europe were the first to discover the horror. They found thousands of children living in obscene conditions in one orphanage after another. Medical teams from France came into western Romania in 1991. Finally, the French doctors arrived at the orphanage on Vicentiu Babes Street in the city of Arad. Someone drew a sample of blood from Ana's tiny arm. The positive test result revealed the worst possible news. HIV, Human Immunodeficiency Virus (or CIDA, as it is known in Romania) was alive and thriving in her body. Her fate had been sealed with an injection of a tiny amount of contaminated blood. From that moment, the system considered her to be "irredeemable."

Hundreds of Romanian children were infected with HIV during this time. Most of those children had already been abandoned into the state system. Others were abandoned after a diagnosis was made, because the stigma and emotional stress were a greater burden than most families could carry. Romania is home to over 50% of all pediatric HIV/AIDS cases in all of Europe.

In the spring and summer of 1990, the rest of the Western world began to hear stories about thousands of children who were abandoned and warehoused in Romania. Television crews arrived to document the tragic story. They began to hear the unacceptable explanation that Ceausescu had ordered the people of Romania to have more and more babies. He had offered tax incentives for families to have more children. He had outlawed birth control and punished families who were not having enough children. His goal was to have control over a larger work force and a huge army—all so he could look like a big-time ruler in the eyes of other world leaders.

The television story hit American living rooms like the news of your mother's death. People were immobilized as they watched the story unfold. The images of children living in terrible conditions shocked the world. Who could believe that so many babies could be abandoned in one place? How could this be possible? Who would do such a thing?

Hearts were broken and lives were changed forever by the power of those living room images from a place few people could find on the globe. Over fifteen years later, I still ask groups if they remember seeing the news about the Romanian orphans on the original broadcasts of 20/20. A majority of affirmative hands still reach upward as they begin to relive those images in their hearts.

Meanwhile, the summer of 1990 brought violent storms to the eastern plains of Colorado. On June 6, 1990, a tornado tore through the small town of Limon. Our daughter Cheri and her husband John live in Limon. They discovered the power of nature that day, and their lives were changed forever. The early-summer tornado destroyed their business before it skipped out of town, just missing their home. By Thanksgiving of that year, their grocery store had been rebuilt and they were giving thanks that no one lost their life to the tornado. They were also giving thanks that God had blessed their lives with a son and with the gift of prosperity. Their thoughts and prayers soon turned to those who were less fortunate. Who could forget the desperate images of the orphans in Romania?

By Christmas of 1990, John and Cheri had come to a decision. God had blessed them to be a blessing to others. They discerned that God had another child for their home. But this time, the child would not come by natural means. This child would be born into their home through the loving action of adoption. The adoption would reach around the world to touch the heart of a child in Romania. They would do their part to bring healing to the children Ceausescu left behind.

January 1991 marked the start of the war in the Persian Gulf. The live-action bombs were exploding on every television set around the world. American warplanes were dropping bombs on Baghdad. Saddam Hussein declared war on anything that looked like American activity. Everyone was on the alert for terrorist activity. Airport security tightened. The war went on.

But even a war could not stop the decision to travel to Romania to find a child. On February 15, 1991, Cheri and my wife Sue boarded an airplane bound for Bucharest. We said good-bye at the airport terminal in Denver, not knowing when they would return. They had made arrangements for someone to meet them at the airport in

Bucharest, but they didn't know how to find a child or how to make an adoption happen. They were determined to learn.

In the next four weeks, they traveled across Romania, endured a freezing winter with no heat, ate whatever they could find, and discovered God around every corner. They traveled over the Carpathian Mountains across the plains of Transylvania to the city of Arad. They found their way to the maternity hospital in the center of the city. Arm in arm, they walked into the maternity hospital wondering, "Is this the place? How will we know which child to pick? How can we choose one and not the other?" It seemed impossible, but remember, God was around every corner!

The first child they saw was only two weeks old. It was a little girl who held the traditional Romanian name, Loredana (Lora-Donna). She had been born on February 5th, and they arrived at the maternity hospital on February 23rd. Sue's journal records it this way:

*"We met Loredana today. Her birthday is February 5, 1991.
She is beautiful and bright eyed and has won us over completely.
God willing, she is the newest member of our family."*

Loredana's birth mother came to town from the little village of Curtici. She knew there were already too many children at home to feed. She gave birth and then abandoned the baby at the hospital. By faith, Cheri and Sue knew she was the one.

By early March they had completed the Romanian adoption process and cleared the American Embassy in Bucharest. They arrived home on March 12th—exhausted, hungry, and fulfilled. They were proudly carrying our newest member of our family, Loredana Louise Walker.

It took weeks to hear about all the experiences they shared in Romania. The people they met, the places they went, the things they learned were all overflowing. They tried to tell us about every emotion and every turn in the road.

I remember the moment when Sue first told me about Ana. She was talking about Loredana's birth family and the Social Services family history that had been done for the Romanian court system. The history revealed that Loredana was the tenth child born to the same parents. The document also disclosed the fact that Loredana had an older sibling, who had also been abandoned. The abandoned child was also a girl, and her name was Ana. Loredana was born in February of 1991, and Ana was born two years earlier, in March of the turning-point year, 1989.

Sue and Cheri didn't see Ana while they were in Romania. The Romanian home study that formalized Loredana's adoption included Ana's name. It identified her as a full sibling and listed her as a ward of the state. The attorney handling the case knew that Ana was in the Dystrophic Hospital near the open market. He offered an opportunity for Cheri and Sue to see her but that did not happen. Cheri only had paperwork for one child. The thought of seeing another child and then departing might be too much to take. They chose not to see Ana but God would not let go. Maybe we could seek out some information about her at a later date. Or maybe it was best not to even think about her at all. After all, there were thousands like her, and we couldn't save them all.

Loredana began to thrive. Everyone loved watching her develop and discover. We tried not to talk about Ana, but that didn't last for long. John finally put voice to the thoughts we had all held in secret. We had to find out about Ana. We had to know more. We had to seek her out. After all, she didn't know it, but she had already become a part of our family.

By the grace of God, we discovered a young woman from our area who had served as a volunteer in the orphanages of Arad. She opened her heart to the children as a response to the same television images that originally captured a whole nation. She thought she might have worked with Ana on one of her orphanage assignments in Arad. She made some phone calls from her current assignment working with street children in Bucharest. She called us to share the bad news. It was true—Ana was HIV+. That fact alone made her ineligible for adoption, or "irredeemable" in the eyes of Romanian law.

We knew what we had to do. We had to find her. We had to do whatever we could to help her. We had to let her know that she was loved. We had to let her know that she was a part of our family. No one wanted to say it out loud, but we needed to give her a place to die.

I didn't know how much this little girl was going to change my life. I didn't know that within a period of seven short days my life would never be the same. The journey was just beginning.

Chapter Two

The Journey

People were bringing little children to Jesus to have him
touch them, but the disciples rebuked them.
When Jesus saw this, he was indignant.
He said to them, "Let the little children come to me, and do not hinder
them, for the kingdom of God belongs to such as these.

Mark 10:13-14 NIV

Day One Saturday, January 16, 1993

My day started in Wilmore, Kentucky, of all places. In many ways, Wilmore is my spiritual home. I had traveled from Denver to Wilmore one week earlier. I returned to the campus of Asbury Theological Seminary, where I was participating in a Doctor of Ministry course—another five full days of class work. Wilmore represented a spiritual womb for me. Sue and I spent three years in Wilmore while I was a student the first time at Asbury Seminary. It was the place where our call to serve was enriched and affirmed. It was the place where we continued to discover what God had in store for our lives. I suppose it was appropriate that my "new birth" experience would begin in the place where I had been nurtured in the Christian faith.

For now, I was more interested in getting packed, grabbing a ride, and checking in at Blue Grass Field in Lexington. At 4:30 PM I was sitting on the ground at Blue Grass Field waiting for the Delta commuter plane to get off the ground. I was bound for Atlanta, where I would meet Cheri. We would travel to Frankfurt, stop in Vienna, and then travel on to Bucharest within the next 20 hours. The adventure was about to begin. I jotted a quick note in my journal.

> *I don't think I'm ready to spend the night on the plane*
> *but I'm excited to see Ana and the country.*
> *It will be a better week than the one I just finished ~*
> *It didn't go so well.*
> *The one thing the disciples asked Jesus to teach them ~*
> *teach me to pray ~ Lord, teach me! ~*
> *Trustful self commitment ~*
> *Father, into thy hands I give my spirit,*
> *my family, my ministry, my love, my ...*

I drew pictures in my mind as I imagined the conditions at the orphanage. I wondered what Ana looked like. I had great faith that we would find her, but I also had tremendous anxieties about what we could do after we found her.

We thought she was probably still in the state orphanage system in Arad. But we didn't know how we would actually find her in the sea of four-year-olds. Even if we were in the right place at the right time, how would we identify her?

Cheri and I met at the airport in Atlanta. We checked our bags, took a deep breath, and set off on an adventure that would be life changing for our whole family. We couldn't predict the results, but we knew God was with us as the plane rumbled from the earth with a cargo of faith and hope.

> *Flight 14, Seat 28 C. We will go over Boston, Philly, New York,*
> *Nova Scotia ~ and then over the North Atlantic.*
> *I just looked at Cheri's permit to bring a sample of Ana's blood*
> *back into the US. She is waiting and she doesn't even know it ---*
> *She is our little girl and she doesn't even know we exist ---*
> *this is a strange world --- Cheri is a blessing --- I miss you Suzi ---*

Romans 8:28 --- I believe it is revealed truth,
I just don't always know how it will come true ~
That's the anxiety and the mystery of faith.

One week earlier, Cheri went to her family physician for lessons on how to draw a sample of blood. She practiced the technique in the office and now was equipped with all the necessary supplies to draw a sample from Ana so we could have it tested in the United States. We needed to evaluate her health situation, but we didn't know what difference the results would make in our eventual decisions.

Day Two Sunday, January 17, 1989

It is Sunday, a day to worship. I have been praying all the way across the Atlantic. I have asked God to teach me what I need to learn from this experience. I have been brash enough to ask for success, even though I am not really sure what would constitute a truly successful trip. Will I be satisfied just to find Ana? Will I regret what we discover? Will I know what to do next?

9:30 AM Sunday 1-17-93 Frankfurt ~ still in the air ---
not much sleep ---
30 minutes from being on the ground in Frankfurt, Germany.
It's 3 degrees on the ground.

I was thankful for the sleeping bag I had stuffed in the expandable side pocket of my garment bag. It made a great pillow while I slept on the white marble floor inside the Frankfurt terminal. We had a short layover before boarding the next flight. Somehow a rock-hard floor and a down pillow became the right combination to catch some much-needed sleep. Frankfurt is the busiest airport in the world. In spite of the heavy foot traffic passing by my marble bed, nothing could disturb my sleep.

We landed in Vienna, but the plane only taxied to the outer limits of the tarmac. I watched as three military vehicles approached the plane. About a dozen men and women in olive military uniforms approached. Half of them were very young men, maybe 18 or 19 years old, with automatic weapons held in a ready-for-action position.

They weren't smiling. I could tell that they were very cold, but they were determined to do their duty.

The soldiers boarded and started a systematic search of the entire plane. Every passenger had to show their passport and airline tickets. We were each questioned about our belongings and our seat assignment, and our ultimate destination. A few passengers were questioned extensively, while others just got the once-over. About a third of the bags were opened and searched, while the flight attendants systematically removed the bottom cushions and outer coverings from each seat. (I had always wondered why those seat covers were attached with Velcro.)

By the time they got to our section in coach, everyone was deadly quiet. It was clear these folks meant business. They had been through this drill before. As they approached me, I could see the insignia on the uniform of the solider who was reaching for my bag. They were Romanian soldiers searching a plane that had been grounded in Vienna before it would be allowed to land in Bucharest. It was my first taste of Romanian cultural paranoia. I would learn later that there was a huge difference between "official" hospitality and the loving acceptance we experienced in the homes of Romanian families.

But for right now, I was just trying not to let my heart rate get out of control. A 19-year-old soldier with a very angry adult look on his face was inspecting my passport as though it was a document I had just created in the spacious airplane bathroom. He asked me to confirm my seat assignment and to identify my luggage. Then he looked me over from head to toe with an expression on his face like Gary Cooper at high noon. I didn't mind at all when he moved on to the next passenger.

We finally took off from Vienna. An hour later, we were ready to land at Otopeni Airport on the far north side of Bucharest. From the Delta window, the airport looked drab, dirty, and unfinished. The blue runway light bulbs were all exposed, but it was OK because all the bulbs are broken anyway. In their place were more very young members of the Romanian Army. They were all standing at attention with the ever-present automatic rifle strapped over their shoulders.

Now we are inside the airport. We are herded through a line to fill out a visa form, then into the "Visa Tax" line to pay our $35.00 to get into the country. From there we stand in line at passport control so another solider can inspect our passports looking for evidence of homemade documents. The airport is dim and dirty. Cigarette smoke fills the air with a stale blue haze. The airport smells rancid, like a dry dirt cellar where burlap sacks have been storing potatoes for the winter.

The officer inside the telephone booth that serves as a passport control office is vigorously stamping my passport. He makes some notes on a small pad of printed newsprint paper. He inserts three layers of worn-out blue carbon paper between the pages of another small newsprint tablet. He fills out a few lines on the three-inch-square form, then tears it from the pad and places the scrap of paper in the back of my passport, before giving me one last look of contempt and directing me forward with a flip of his hand. Later I discovered that I would need that scrap of newsprint to get out of the country. I'm glad it stayed in my passport

As the luggage arrives, we fight off the men who want to earn a tip for helping lift a suitcase. There is dirt everywhere and lots of loud conversation about the ownership of certain bags. A dog in a makeshift traveling kennel is barking incessantly—*loudly*— in the corner. The airport is cold, the lights are dim, and the atmosphere is like a bad Nazi movie from 1949.

Mike (his name is the American version of Mihai) meets us as we come through the hoard of taxi drivers at the end of the ramp that leads into the main terminal. Fortunately for me, Mike speaks English very well. He even knows how to use American four-letter words in the right context. He helped Cheri and Sue during their trip in 1991. Mike is a student at the school of law in Bucharest. In a couple of years he will graduate and become an "advocate" with a specialty in helping people reclaim title to their property that was taken by the government during the communist era.

Mike takes us to the apartment where his mother and father live. The streets are dark and narrow. The cobblestone streets bounce Mike's Dacia (the Romanian national car) until it feels like we are out of control. It is a wild ride through the unforgiving streets of

Bucharest. Mike honks his horn at every motorist and pedestrian in sight. Every other driver seems to love using his or her horn just as much. I asked Mike why everyone honks their horn so much when they drive. His response was quick and informative, "If I don't horn, how will they know I am coming?"

We arrive to meet Mike's parents. Neither speaks English, but Momma apparently lives to cook. She fixes us potatoes and "the meat of the pork" before we head off to get some sleep. Cheri will stay here, while I will sleep on the couch at Mike's apartment a short drive away.

Mike's apartment is stark. There are no pictures on the walls or decorations to brighten the dull white stucco interior walls. The furnishings are sparse. A table in the kitchen is large enough for two. He keeps three small stools stored under the table for seating. There are closed doors everywhere to keep the heat localized. Mike turns on the television to reveal a green glow. Faint images of a soccer match (fotbal) move across the screen. He apologizes but explains that television is much better, now that the government does not control it. During the Ceausescu days, there was only one station available. The 24-hour broadcasts included reruns of Ceausescu's speeches and propaganda about the prosperity of Communism. He offers me the couch, a blanket, and a pillow. I am thankful for the opportunity to rest. My brain is beginning to feel a little fuzzy from all the travel with no sleep.

Chapter Three

The Breakthrough

For this reason I kneel before the Father,
from whom his whole family in
heaven and on earth derives its name.
I pray that out of his glorious riches he may strengthen you with power
through his Spirit in your inner being, so that Christ may dwell in your
hearts through faith.

Ephesians 3:14-17 NIV

Day Three Monday January 18, 1993

5:30 a.m. up and ready to go to the airport again to head for Arad.
7:30, on a Tarom "airliner" AN-24RV ~ it looks old.
We meet Dawnalyn at the airport.
Mike has taken care of everything.
We would be lost without an advocate.
I want to take photos of the airplane for a pilot friend back home,
but the kid with the automatic weapon
doesn't think that's a good idea.
We are on our way to Arad with hope that we can get in to see Ana
Note: Mike says the church during the Ceausescu years was 'low'
He heard that five churches in Bucharest were destroyed
but they were always open.
Now everything is OK --- according to Mike --
but still not strong. He got a Bible that we sent earlier from the church.
He thinks people are thankful for Bibles.

There is so much vibration on the plane --- it is hard to think. Cheri and Dawnalyn are yelling at each other in Vibrato. Dawnalyn is a young woman from the Denver area. She has been working as a volunteer in Romania since just after the Revolution in 1989. She worked with Ana in the orphanage when Ana was very young. She has a photo of Ana sitting on the floor playing with a ball. Dawnalyn will travel with us to help open some doors with the director at the orphanage and to help us identify Ana when we find her.

The domestic airport in Bucharest is worse than Otopeni International. Groups of women are selling small black seeds along the sidewalk. They use a little wooden cup to scoop a handful of seeds from a bundled scarf. Sales are not too brisk this early in the morning.

Four ragged street children approach us to beg for money. They extend the palms of their hands with big eyes and dirty faces asking for coins. Mike chases them away with a loud voice to protect his clients from the harassment of gypsy street children. He gives us a quick instruction in street survival skills: "No matter what, don't look at them and don't give them money." He knows we will have many opportunities to deal with the children of the street.

I have to duck severely to enter the Tarom airliner that will take us to Arad. The seats are tattered and very small. I am crammed into a small space that has no place for my legs, and I cannot put my knees straight in front of me. I knock the wooden tray table from the back of the seat in front of me as I try to squiggle into the space allotted. The propellers start to turn slowly as the engines begin to roar. The pilot and co-pilot are sitting on the other side of a dirty maroon curtain as they go through their checklist. I begin to pray as we leave the ground. The plane is vibrating like my old '58 Rambler. Apparently there is no such thing as a non-smoking flight. All the passengers are smoking. Even the flight attendant is smoking. I am afraid to look out the window for fear the plane is smoking, too. My thoughts turn to Ana as I make another entry into my journal.

John 1:35-50, Jesus calling the disciples ~
From Gordon McDonald ~ How absurd that we would call ourselves
disciples yet do nothing he says to do.
I wonder what the world would be

if the church was the "church" and disciples were "disciples."
Lord, help me to see, and understand, and act

Eva Jurca (the wife of Cheri's attorney) picked us up at the airport in Arad. She was wearing a fur coat that was far above the norm for Romanian winter outerwear. We took a quick tour of the city before stopping at her store. She was one of the first entrepreneurs in the city since the revolution made free enterprise possible. Her store is a 10×10 foot freestanding building that looks like a log cabin. It sits at the end of the main street in Arad, Revolution Boulevard. The tiny building looked out of place among the ancient architecture of the city. Inside, the store had counters around three walls, with just enough room to turn around in the center of the floor. She sells a few shoes, some cosmetics, a few bottles of wine, kilo-size bags of rice, laundry soap, and assorted clothing from a rack in the corner. The smell of bad perfume lingered with us for hours after we leave the shop.

Finally, we are in Arad, Romania. We will stay with Sever and Eva Jurca while we are here. Cheri and I will share a tiny bedroom that has Michael Jackson and Axel Rose posters on the walls. The room belongs to the Jurca's pre-teen daughter Elena. I get the floor. Cheri gets the bed, which is the size of a small couch. The apartment is not spartan like Mike's place in Bucharest. The large bookcase in the living/dining room is packed with reading material of all sizes and shapes. The main room and the two bedrooms all have the feel of being over-furnished. There is hardly a place to turn around after you enter a room.

After a quick lunch, we make our way on to the cold, diesel-fumed streets of Arad. We walk to the Vicentiu Babes orphanage for our first meeting with the director, Dr. Lia Laios. Cheri remembers the way to the orphanage from her first trip here in 1991. This is why we came.

We walk past the free market on the corner, where people are selling everything from cigarettes to "fresh" pork and poultry. The market leads us down the street to the little white triangle-shaped flower shop, then one block more to the orphanage. My mind is full of questions. Is Ana here? Will we be able to see her? Is she still alive?

The ancient orphanage building sits behind a few large trees. A tall iron fence sits on top of a short brick wall. The iron entry gate looks foreboding. The iron bars have obviously been painted many times. The layers of dusty tan heavy gloss paint are chipped in places to reveal generations of history. I wonder if Ana entered through this same gate. I wonder how many children have entered here.

We arrive with "gifts" of candy and cigarette lighters to say thank you to the guard who oversees the gate. He is sitting in a small guardhouse. He is dressed in street clothes but he is wearing a short blue smock over his dirty jeans. We ask to see the director. The guard gives us the once over. He knows we will give him something on the way out if he helps us at this point. He nods and gestures for us to follow.

Cheri, Dawnalyn, and I walk up the 20 or so steps and enter the thick walls of the building through the 10-foot-tall entrance. Two hundred and fifty infants and young toddlers live behind these ancient looking walls. There are no children in sight, and there are none to be heard.

The place is cold and mysteriously quiet. The ceilings must be 12 feet high, with arches over every doorway. The walls are painted concrete stucco. They are light yellow gloss about 4 feet from the floor. The upper wall is painted a chalky tan color that seems to absorb all the light from the large, curtained windows. There is a hand-brushed thin line of dark brown paint that separates the upper and lower portions of the walls. The whole place needed to be repainted and repaired several years ago.

The black and white squares on the floor lead us to the director's office, where the guard announces our presence and then points to the small couch in the hall. We wait there and wonder what might be around the corner. Is she here? Will we find her? Is she still alive? Imagination can bring hope or it can become captured with grief that knows no limit.

After a short wait, Dr. Lia Laios steps into the hallway and invites us to join her in the office. She is the Director of the Leaganul Pentru Copii (The Organization For Children). She speaks a very small amount of English. We are relying on Dawnalyn to translate and communicate on our behalf.

Dr. Laios remembers Dawnalyn from last year, when she served as a volunteer in the orphanage. She is happy to see the photos Dawnalyn brought. One by one, they talk about the children in the pictures until Dawnalyn finally turns to the photo of Ana sitting on the floor. "Do you remember Ana?" "Da, da, Ana CIDA!" She remembers! She knows where Ana is! Dr. Laios has authority over another orphanage where children go when they get too old for Vicentiu Babes. She gives us permission to look for Ana there. She even calls ahead to give her approval for us to look for Ana.

A brisk 15-minute walk brings us to the front door of the orphanage called Tudor Vladimirescu. We discover the secret to the apparently locked gate and make our way into the yard. The building looks like a 1950s two-story American elementary school. It is exposed red brick with a very square look. Nothing fancy. Just your basic building.

We are filled with anticipation as we ring the bell. We do our best to introduce ourselves to the nurse at the door. We tell her that Dr. Laios sent us. The response is not warm and welcome. The nurse says it will not be possible to enter—the children are all sleeping. It is 3:00 in the afternoon, and it is naptime for the sixty-four kids who live here.

I think to myself, surely they can't all be asleep. That was my ignorance of institutional life showing. The fact is, when it is naptime, it is naptime for everyone. Obedience, complacency, and lack of individual identity are the primary colors of this institutional lifestyle.

Another nurse comes to the door. Her nametag says "Rodica." She speaks a little English. We make our case once again. We have permission from Dr. Laios. We are looking for Ana Cionca. Do you know Ana?

We are sorry that it is naptime but we have no time to waste. We must find her. We must look for her. Can't you make an exception? We really need your help. She is beginning to soften. "Multumesc," thank you, thank you, we will be very quiet. She reluctantly holds the door open. We are in but we don't know what to do next. We ask again if they know Ana?

The Romanian institutional culture doesn't lend itself to answering questions that may get you in trouble with a superior. They choose not to answer as they direct us into the first room just to the right of the entryway.

The room is stark and cold. There is a strange but familiar mix of odors in the air. It smells like stale urine covered by the very faint odor of bleach. There are sixteen white iron cribs in the room. Each crib is only three feet long and about two feet wide. All sixteen are occupied. There are four iron cribs along the outer wall. Eight cribs are back-to-back down the center of the room, and another four are lined up along the wall with no windows. The children are all three to five years old. And they are all sleeping—each one covered with a single tiny blanket.

We approach the first crib with a prayer in our hearts. Cheri lifts the blanket to expose a little face. Dawnalyn thinks she will know Ana when she sees her. She takes a look and shakes her head. It's not Ana.

The nurses in the doorway are whispering to each other. They are distressed because we are going to disturb naptime. We move three steps to the right and assemble with the same prayer at the next little white iron crib. Cheri lifts the blanket, Dawnalyn shakes her head, and we move on to the next crib. The whispering in the doorway is getting louder.

There is really no way to tell the boys from the girls. Most of the children have dark hair and dark eyes. There are a few blondes and an occasional redhead, but they all look the same. They all have the same institutional haircut. A few of the children are bald from having their heads shaved. I recognize the tell tale sign of lice from my days working with mentally ill children.

We continue to go from crib to crib. Each time Dawnalyn looks with all the energy her memory can muster. Each time she shakes her head. A spirit of discouragement is beginning to pierce the edge of our hopefulness as we finish looking at the first sixteen children.

We move to the next room just down the hall. The door to the room has frosted glass from top to bottom. The glass rattles in the frame as the door closes behind us. There are sixteen more children in this room. The layout of the room is exactly the same. There were four cribs along each outer wall with eight back-to-back cribs making up the center section. We start the process all over again. We lift one blanket at a time. We look at one dark-eyed child after another.

Three nurses are following us. This time they have gathered in the corner of the room to whisper. We don't know what they are saying, but the atmosphere is getting more stressed. They think we are intentionally waking the children from their naps. We are breaking the schedule and upsetting the institutional clock. We are making their job more difficult.

Dawnalyn shakes her head repeatedly as we engage one child after another. Maybe Dr. Laios really didn't remember Ana. Maybe she really didn't know where she was. Maybe we really are just disrupting the institutional schedule without any real hope of discovering the child we have traveled around the world to find.

The nurses are getting more outwardly angry and anxious as we ask to go upstairs to see the other rooms. We make our way up the stairs with them following even though they would rather we leave their kids alone and get out of the way. Upstairs we find sixteen more sleeping children. Sixteen more blankets. Sixteen more times Dawnalyn says no, this is not Ana.

Finally we are standing at the entry to the final room. Without opening the frosted glass door, I already know the layout of the room. Four cribs on each side with eight down the middle. I want to go in but I am afraid we will come up empty. Rodica suggests that it is time for us to leave so the children can sleep. We don't want to be difficult, but we have to look.

I open the door. I am right; there are sixteen more white iron cribs with sixteen more children. But something is different here. Only fifteen of the kids are sleeping beneath their tiny blankets. One child is standing up in the crib. Every eye is drawn to the standing child. Dawnalyn almost shouts, "It's her! It's Ana!"

We ask the nurses, "Is this Ana Cionca?" Moments ago they didn't know any child by that name. Now they give us an affirmative nod. They knew where she was all the time. It is Ana. We have found her. Out of sixty-four children, she is the only one awake. She is standing there looking right through us like she had been waiting for something to happen.

Her hair was crudely cropped and matted in the institutional style of the day. Her eyes were dull. She didn't smile. She made no effort at eye contact. She just stood there. She was so little and so thin. She was wearing several layers of clothing. I could see a shirt that was covered by a sweatshirt tucked tightly under her worn green corduroy overalls. Her crib had years of heavy white paint chipped away in layers to expose the cold dark iron frame underneath. I'm guessing the layers of white paint on the crib are all lead based—poisonous to children.

Her fingers were gripped tightly around the chipped rail. They were so thin, fragile, and almost transparent. She looked the part of a waif in Oliver Twist, but she wasn't standing up to beg for more food. She wasn't outwardly asking for anything. She wasn't reaching out to be held. She wasn't smiling. She wasn't making eye contact. She didn't know and she didn't care that we came halfway around the world just to find her. She just stood there and looked through us into the empty space that was her world.

We saw her for about two minutes before Rodica insisted that we must go. Tomorrow? We asked. Can we come back tomorrow? She gave us a nod that seemed to mean, I will open the door if you don't tell anyone. I saw the anger and distress depart from Rodica's face. Her eyes were filled with tears. She saw our need, she knew Ana's pain, and she wanted to help. In the days ahead we would find the same response repeated by other nurses. The outer hardness, distance, and anger would melt to reveal warm and compassionate hearts.

The door to Ana's room rattled closed behind us. We turned the corner and walked down the main staircase to the front door. The air was crisp and cold. Somehow it seemed cleaner outside. The sky seemed to change from gray to blue. We were smiling through eyes that were filled to the brim. We did it! We found her! It really was her! We talked and smiled so much that we didn't even notice the bitter January cold as we walked back to the apartment for the evening. It was a great day. My journal documented our excitement as a poem came to my heart.

Slept little through the night but I am thankful for the sleeping bag.
At 2:00 in the morning I looked at the clock. Cheri was awake too.
We talked most of the night.
Me on the floor, her on the bed above me
We were careful not to wake Sever and Eve.
I just kept thinking about her face.
I can't believe we actually found her.
Thanks Suzi for your prayers.
I want to talk with you. I can't stop thinking about those kids.

Sixteen Children

16 children, all in their row.
16 children, does anyone know?
16 children, behind bars of white,
16 children, who will tuck you in tonight?

16 children, with blankets over their heads.
16 children, will you ever have a chance to wed?
16 children, obedient, blank, stiff and sore,
16 children, can your pain increase any more?

16 children, it seems so easy.
16 children, your food makes me queasy
16 children, you need a friend.
16 children, it must come to an end.

16 children, the Lord loves you.
16 children, but how? what shall we do.
16 children, you are 2, 4, and more.
16 children, the world must come the Lord.

16 children, I will rock you in my dreams.
16 children, my tears fall at your screams.
16 children, I pray for you ~ peace.
16 children, for under the cover is a beautiful face.

For Ana January 1993 Arad

Day Four Tuesday, January 19, 1993

We were awake early the next morning. I don't think we really slept at all. There was so much to think about and so many things we needed to do in such a short time. Eva fixed a wonderful cup of Romanian coffee for my breakfast. She thinks it is strange that I drink coffee without sugar. Their standard coffee preparation calls for thick, dark Turkish coffee that is converted to syrup by the sugar that is stirred in. I like mine straight up and plenty black. This cup will take care of any caffeine I might need for the entire day.

We create a plan for the day as we eat a breakfast of white goat cheese, sliced salami, heavy Romanian bread, and the incredible coffee. We want to be at the orphanage to talk with Dr. Laios when she arrives at 9:00 AM. Our goal is to convince her to allow us to draw a blood sample from Ana so we can have it tested when we arrive home. Cheri has prepared for the blood draw with a one-day lesson from her family physician. She also made advance arrangements for a blood work-up at Emery University in Atlanta. Atlanta is the first place we will step back on American soil. We want the testing to be done without any more delays than necessary.

Dr. Laios was just taking off her coat and hanging it on the hook in the corner near her desk as we arrived at her office door. She invited us to sit around the dark conference table that took up most of the office. She demonstrated her hospitality by asking the secretary in the outer office to get us some coffee. As she closed the oversized office door, I noticed that it was upholstered from top to bottom in padded black leather-like material with decorative tacks, creating a diamond pattern that meets in the center of the door. It is a distinctive office door.

We had just started the discussion by looking at some photos of Loredana when the coffee arrived. It came in a tiny espresso cup with sugar already generously installed. I thanked her, drank all but the bottom accumulation, and declared it be wonderful. It was actually stronger than Eva's breakfast coffee!

The photos and conversation about Loredana were very important to our discussion. We wanted Dr. Laios to know that Loredana was strong and healthy. It not only provided a full sibling connection to Ana, but it also helped reverse the mythology about foreign adoptions.

Rumors had circulated throughout the country about what happens to Romanian children who are adopted by foreign families.

The rumors had it that children were adopted so their organs could be harvested for transplants, or so they could be put to work in a form of slavery. Some people thought the Romanian government was responsible for the rumors, which had started back in the summer of 1991. They may have needed a reason to shut down the adoptions that were happening immediately after the revolution so they could save face with the rest of the world. Photos of happy, healthy children were an important part of closing down the rumor mill.

We told Dr. Laios that we had found Ana. We were very interested in helping Ana, and were even interested in the possibility of adoption. Dr. Laios looked very serious when she told us that adoption was not possible because Ana had tested positive for HIV. Only children in good health can be considered for adoption. Ana's health situation placed her in the "irredeemable" and non-adoptable category. We assured her that we understood the situation. We just wanted to confirm Ana's health status by drawing a blood sample to be tested in the United States.

Dr. Laios was adamant. It was not possible to draw blood on this child. Even if it was possible, she was not in a position to authorize such an unorthodox request. If we wanted to pursue it further, we would need to go back to Bucharest to seek permission from the Ministry of Health. Her position was clear. She did not have the authority to give permission for the blood draw, but she did give us permission to see Ana at 2:00 that afternoon. We could spend some time with her, under the condition that we do not take any photos of Ana or the other children. We assured her that we understood her situation and that we appreciated her willingness to help us. We promised to be back in touch with her when we had obtained documentation from the Ministry of Health in Bucharest.

The five-minute walk back to the apartment left us cold to the bone. The temperature had dropped and the wind was whipping through the streets. We began to thaw out as we sat down at Eva's table for a quick lunch of chicken soup and some sliced oranges. Eva had found the oranges at the market and bought them as a treat for us. They were good, but they would never have made it to a produce counter in the United States. They were small, bruised, and shriveled. But she found them for us, and she stood in line over an hour just to buy them. She was pleased that she could provide us a fresh orange in the middle of the winter. We were grateful for her hospitality.

The New Testament Connection

I brought a small duffel bag full of paperback Romanian New Testaments from the International Bible Society with me to give away. So far, I had given copies to friends and people we met along the way. I still had about 45 New Testaments, and I was running out of time. I asked Eva if there was a church or someplace where I could distribute the Bibles. She didn't really know, but she thought the building about one block away was a church of some kind.

Eva and Sever, like 96% of Romanians, considered themselves to be Romanian Orthodox, but they had not been to the church since an Orthodox priest married them. She opened the window shade and pointed down the street to the building that she thought might be a church. If I was going to give the Bibles away, I needed to do it right away. We didn't know if we would have time later.

I pull my coat on and make my way down the street. By this time, the wind is still blowing and it has now started to snow. The building Eva pointed to doesn't look like any church I have ever been in before. I can read enough Romanian to know the sign on the front of the building identifies it as a Baptist congregation. A gray iron fence separates the front of the building from the street. I try three different doors before I find one that is open, all the way in the back of the building. The wind is blowing the snow in my face as I open the large wooden door to enter the church building.

I enter only to discover that it is almost as cold inside as it is out in the snow. I hear singing coming from upstairs, so I make my way toward the music. I open another large, wooden door to discover a huge sanctuary. It has a high ceiling with ornate ceramic chandeliers hanging from the center. It looks like individual theatre-style seating for about 1,000 people. The singing I heard is coming from the choir loft, where 60 to 70 high-school-aged students are having choir practice. They become silent at my intrusion into their practice time.

The man who has been directing the group lays down his baton and takes a few steps my way. I ask if anyone speaks English and I am pleased with his positive response. He greets me warmly, even though he is wearing a coat, hat, and gloves to direct the choir. The students are also wearing coats and scarves. I tell the choir director that I am from Colorado and that I have some New Testaments. I ask if he thinks the church could use them. He responds with a huge smile and says, "My students can use them."

Now there is only one problem. I need to go back to the apartment to get the New Testaments. It will be no problem to the choir. They will wait. I make a quick dash back to the apartment through the snow. It is only a short distance but the wind is bitter and the snow is accumulating. I grab the bag of Bibles and make my way back to the church.

As I re-enter the sanctuary, I am surprised as the students stand and applaud my arrival. Apparently, their director has told them that I have Bibles for them. They are all smiles, and I am a bit embarrassed because I don't have enough to give one to every student. As we distribute the Bibles through the choir loft, I discover that several of the students speak wonderful English. They thank me over and over again. The director tells me that this is the first time these students ever had a Bible of their own. As I apologize to the group for not having enough, one young man in the back row stands to speak for the whole choir. He calls out in a clear voice, "We know how to share."

The director invited me to speak to the students. He says they are interested in knowing what American teenagers are like. He will translate my remarks, although many of the students understand English. I tell them I really don't know very much about Romanian teenagers, but I imagined that American teenagers were about the same. They enjoy their friends. They like to listen to music. And they love to go to school. The last comment brings smiles and the expected negative teasing response.

I ask them what they think American teenagers are like. One young man in the back stands to speak for the group "We think

American students are like Michael Jackson and Axel Rose." We all laugh as I remember the posters that decorate the bedroom where Cheri and I slept. I ask the students if they know anyone like Michael Jackson or Axel Rose. They laugh and reply that they don't know anyone like that. I assure them that I didn't know anyone like Michael Jackson or Axel Rose either.

This fun encounter would have a lasting impact on my relationship with the church in Romania. Years later, many of these same students would become close friends and coworkers for Christ.

Before I departed, the director and I exchanged information. His name is Ioan Redis and he is the Principal of the Christian High School in Arad. He proudly told me that the students are the first group to take part in the first private Christian High School in all of Romania. He was excited about bringing the possibility of a Christian education to the youth of Arad. We were together in that huge frozen sanctuary for only a few minutes, but our lives would become deeply connected. Little did I know that Ioan Redis would become one of my dearest friends.

To the Church:
It is true that God speaks to us in many ways.
The Bible speaks, our personal experience speaks,
the witness of others speaks,
and even creation speaks a word from God.
But how often do we listen?
How often do we expose ourselves to the intentional Word of God?

Do you have a Bible in your home?
We naturally assume that the answer is, yes.
But there are more important questions.
The most important question is,
"Do you read and study the Bible that is in your home?"

You may not know it, but you sent me to Romania
with some special gifts.
You sent a case of Bibles that I carried in my suitcase.
They were heavy in more ways than one.
I gave those Bibles to friends and to a group of students
at a Christian High School. They were overjoyed.
For 25 years, Bibles were illegal in Romania.

Even in the last year it was illegal to bring more
than one personal copy of the Bible into the country.

Those students were filled with joy because
they now have their own personal copy of God's Word.
It was impossible until just one year ago. Now it is legal,
but Bibles are far too expensive for the average person to purchase.

Today those students will use those Bibles in their classes.
They will carry them from home to school and back again.
They are a prized possession to be read over and over again.

Perhaps we should do the same.

Back to the Orphanage

The cold and the tight schedule left little time for enjoying new friendships. Our primary goal that day was to spend more time with Ana. We needed to head back out into the cold right away. We didn't want to miss our 2:00 time to see Ana. As we made our way through the streets of Arad, we passed a little shop where someone was cooking pork skewered on a stick. The pork smelled much better than the diesel fumes that seemed to hang everywhere.

Now we discover a "back door" to the orphanage building where Ana has apparently been living for the last year. It is a large double door that opens directly off the street. It has a hand-made latch with an over-sized keyhole that hasn't been used for years. At one time it must have been painted green.

Both doors swing open to reveal a small alley that gives access to three homes and the back of the orphanage. There are small garden plots and grape arbors that are covered with dormant vines. The houses have a "built by hand" look to them.

As we approach the back of the orphanage, I can't help but notice a wagon that is parked there. A single horse is hitched to the wagon. The horse looks worn out and exhausted. There are two large silver barrels in the back of the wagon. The driver is nowhere to been seen, but it is apparent that there is some connection between the wagon and the orphanage. As we ring the bell on the back door, we encounter the driver of the wagon. He has come to talk with the cooks. He is there in the morning to take their order for flour cornmeal and whatever they may need. He will return in the afternoon to deliver the supplies. The sound of his old horse shuffling on cobblestone will soon become a common rhythm to our ears.

One of the cooks comes to the door to answer the bell. We ask for the director while she wipes her hands on her well-worn white apron. She unlatches the door and invites us in with a smile. "Poftim," she says as she points the way down the hall. The first door to the right is the kitchen. I take a quick look to see a large woman in an apron stirring a huge pot of something that looks like Cream of Wheat. Sliced cheese and salami is on the counter. They must be preparing lunch for the sixty-four kids who call this home.

We are lucky; Rodica is on duty again today. She knows what we want, so it saves a lot of time trying to communicate with little knowledge of the Romanian language. She opens the door to a room and invites us in with a gesture of her hand. The door has a sign that says "Medic Clinic." We assume it is an examination room. We later learn that the room serves multiple purposes. It is an office, a staff lounge, an examination room, and a visiting room for the occasional person who stops by to see a child.

Waiting in that room, we didn't know quite what to expect. Maybe we were waiting to get permission from someone else. Perhaps Rodica just wanted to get us out of the way so we wouldn't disturb the children like yesterday. We were overjoyed when Rodica arrived a few minutes later with Ana in tow.

Rodica held Ana's hand as they entered the room. There was still no real expression on her face other than a mild look of confusion. She must have been wondering who these people were and what was going to happen to her now. She wobbled into the room.

Even though she was going to be four years old in one month, she was very unsteady on her feet. She walked with a great deal of difficulty. Her feet were spread apart to give her a better sense of balance. Her knees didn't bend and her hips would only allow her to take very short steps. Without Rodica's hand she wouldn't be able to stand for long. Someone had tried to comb her hair to make her presentable for our visit. She really did look like a waif.

There we were—Cheri, Dawnalyn, and me—alone with Ana. We had not only found her, now we had a chance to get to know this little girl whom we had all been seeking in our hearts. At first she just stood there, leaning on the closest person. Her lower lip was almost raw from biting it with her top teeth. She didn't smile or show much sign of anything going on around her. Slowly she began to warm to the situation. It was like watching a slow-motion film of a flower blooming in the spring. She began to make eye contact. Her face began to relax. Her body language was more open. She even began to smile just a bit.

At that point in her life Ana had no verbal language. She did make a quiet squeaky noise with her lips. Except for an occasional sound that would count as a smile that transitioned to a tiny giggle, there were no other sounds. We talked to her, rubbed her back, and tried to estimate her weight through three or four layers of clothing. She was slowly responding. We wanted more than anything to take a photo. We needed some record so we would have something to take back home. We didn't know if we would be able to see her again, but we had been forbidden from taking pictures.

Occasionally one of the staff members would stop by. The first one spoke to us and tried to warn us that she was HIV+. They were all afraid that we didn't know what we were doing. They were trying to take care of us. They wanted to warn us so we would not become infected through physical contact with Ana.

We started to get another glimpse into her world. It was a life of isolation. Minimal care was offered out of fear of contamination. No one talked to her, played with her, or picked her up out of her crib. She was alone in a world where other children surrounded her.

Every time we got brave enough to try a photo, someone would come down the hall or stop by the room. We didn't want to lose our chance of seeing her the next day by breaking the rules. But finally we took a chance. Cheri held Ana in her lap while Dawnalyn stood watch at the door. I was in charge of the camera. Just as I was set to take the shot, Dawnalyn jumped in to the doorway with a look of fear on her face. Rodica was coming. It was too late. The flash went off and I quickly stashed the camera. Rodica must have seen the flash. She knew what we were doing. She walked past our visiting area and disappeared into the next room. We never heard a word about that illegal photo that recorded Cheri, Ana, and Dawnalyn halfway into the room with her arms in the air.

Our time with Ana that afternoon was a gift from God. We were able to hold her, talk to her, play with her, and begin the process of letting her know that she was loved. It was a wonderful time together that came to an end when one of the nurses came to take her back to her room because it was time to eat. At that point we began to learn that Ana had never eaten solid food. She had spent her entire life eating liquids. First she ate the basic gruel-looking orphanage formula from a bottle. Later we would learn that she had been staying alive even though she was only fed tea with a biscuit dissolved in the water. When food is short, the stronger kids get the best food. None of the food was really wonderful or nutritious, but if you are "irredeemable" there is no need wasting food.

On the way out of the orphanage that day, one of the nurses greeted us on the staircase. She spoke fluent English. When she learned that we were from Colorado she mentioned the name of a man who had been to the orphanage to adopt one of the children. We knew the man and his child from other adoption connections. She was excited to be able to send a message back to Colorado to continue the relationship. She also took us under her wing and gave us a short tour of the rest of the facility.

We went into a little room downstairs where she proudly showed us how the children were being taught to care for their own hygiene needs. They were being potty trained. The room was about six feet wide and ten feet long. There were sixteen children being potty trained in that small space. A 2×12-inch plank had been placed as a "bench" along each side of the room. Eight kid-sized outhouse holes were cut into each plank. Concrete blocks held the planks at the right height so a little potty pan could slip in under each hole.

There they were. Eight little faces sitting on the right and eight little faces sitting on the left. We couldn't tell how long they had been sitting there, but the attendant was proud to tell us that they were not only being potty trained but they were learning how to go to the bathroom at the right time of day. Each room of sixteen children would have their turn sitting in the potty room twice each day.

The Eyes of God

Around the corner, just down the block.
Behind the school, the gate is locked.
Inside the door are the echoes of quiet
If the world could see, it would create a riot.

Eight holes in a long potty plank.
Eight on each side ~ but the eyes are blank.
They obediently sit, some eyes look our way
One thing is clear ~ they must not stay.

Not hopeless ~ but still not clear
How will the gate open for eyes so dear?
Can it be the idea of the God of Light?
We must become his heart to end this plight.

Salvation for some is the warmth of heart,
For others it is the newness of a clean fresh start.
But for these eyes of quiet the need is very clear.
Salvation is the arm of God with a hug for my dear.

The walk back to the Sever and Eve's apartment gave us time to talk about the day and to process the experience. It was a mix of joy, bewilderment, discovery, and frustration. I was tired from the travel, exhausted from the emotion, and done in by the dilemma these children faced. On the surface it looked like a situation that could be fixed. But there were so many children. Right now we had to find a way to fly to Bucharest first thing in the morning, so we could go to the Ministry of Health to plead our case for drawing a blood sample from Ana.

Supper was interesting. We had wonderful chicken broth soup that hit the spot on a cold January evening. The soup alone would have been a fine meal, but there was more. Eva cleared the soup bowls and then presented small sausages that had been fried. These were eaten with cheese and bread. They were spicy but very good. Eva kept letting me know that there was more and that I would want to save space for the real meal.

Finally the main course arrived at the table. It was a heaping platter of boiled chicken livers. Eva beamed and Sever was very pleased to be able to offer us such a fine meal. I just looked at Cheri, smiled and put some chicken livers on my plate. An abundance of Romanian bread and sparkling mineral water helped the medicine go down.

We told Sever about our situation and our need to go to Bucharest. We asked his advice as an attorney. Should we go or is there another way to accomplish the same thing without traveling to Bucharest? He was adamant—there was no other way. We must have the appropriate papers with the right signature and the proper stamp from the Ministry of Health in Bucharest. He suggested the 7:00 AM flight so we would have time to get to Bucharest, spend time at the Ministry of Health, and perhaps still have a chance to get back to Arad the same evening. It seemed like a long trip and a large expenditure. We were running out of time.

God chose that moment to teach about miracles. Just as we had resigned ourselves to taking an early morning flight to Bucharest, God sent a solution. It was delivered in the form of a large man who came by the apartment to check on Sever. He was Sever's physician, Dr. Doru Lazar. Dr. Lazar came to the apartment to administer an injection of medication to Sever. He sat at the table eating chicken livers like they were popcorn as he and Sever talked about soccer. Their laughter and warm conversation was the kind that is only shared among close friends.

As Dr. Lazar ate chicken livers, Sever told him about our situation. He listened patiently. His demeanor turned from the interaction of friends to the professional concern of a physician. He asked us a few questions before revealing that he was the chief of pediatric medicine for the County of Arad. He recognized Ana's name because he was responsible for tracking children who are HIV+. His recommendation was that we did not need to go to the Ministry of Health in Bucharest. We explained that Dr. Laios wanted a document authorizing the blood draw. Without the proper documentation, she would never allow us to draw the blood we needed for stateside testing. He popped another chicken liver and looked rather gruff as he spoke. "I will take care of the paper. Will 9:00 in the morning be soon enough?"

"You can do that?"

"Yes, I can do that." He got up from the table, shook hands all around, thanked Eva for the meal, reached back for one more chicken liver, and disappeared into the cold and darkness of the night.

Day Five Wednesday, January 20, 1993

At 9:00 the next morning, we had the document in our hands. It gave us permission to draw blood for testing. It had signatures, stamps, and official-looking seals. We were excited as we approached the Vicentiu Babes orphanage to meet with Dr. Laios. We were sure she would recognize Dr. Lazar's signature and his authorization.

Dr. Laios looked at our document as we all sat around the conference table in her office. She read it twice before declaring that this paper would not be enough to gain her permission. She said the paper meant nothing to her. We would still need a document from the Ministry of Health in Bucharest. We asked her to reconsider and told her about our conversation with Dr. Lazar. She would not relent. She would not take responsibility for this decision. But she did consent to consulting with another doctor who would be coming to the orphanage in the next couple of hours. We thanked her and settled in to wait outside her office.

The little couch in the hallway has become familiar territory as we wait for another opinion. The morning is almost gone by the time he arrives. We feel fortunate because he speaks English and is willing to hear our story. We cover all the details and then learn that he also knows Ana. He doesn't see a problem and tells Dr. Laios that he will take responsibility for the decision. He signs off on Dr. Lazar's document and seals his signature with a little rubber stamp that he carries in his pocket.

Dr. Laios looks relieved. Now it feels like she wanted to authorize the procedure all along. She just didn't want to be responsible if something went wrong. Years of Communist oppression will do that to people. No one wants to get in trouble or make the wrong decision. The key is to never make a decision. Always pass it along to someone else so they will be the ones to get in trouble. The end result is a bureaucracy that becomes absolutely impossible to deal with. Simple decisions never get made, or they take months, with mounds of paperwork, signatures, stamps, and seals.

Now that someone else has taken responsibility, Dr. Laios is more than happy to help us. We could now go to see Ana to draw the blood. She would call ahead to pave the way. We could even take pictures of Ana as long as we didn't take photos of any other children. She even assigns a lab technician to go with us to draw the blood. The technician is Daniela Motocan. She speaks a little English but not enough to have a conversation. She walks with us to the Tudor Vladimirescu orphanage where she will meet Ana for the first time.

We are escorted into the same examination room were we visited with Ana the day before. Nothing has changed. No one has picked up the dirty laundry in the corner or taken time to sweep the floor. As we wait for Ana to arrive, Cheri shows Daniela the equipment to draw the blood. Daniela has never seen this kind of syringe blood-draw system. She is accustomed to a needle that provides an open drip into a test tube. But we need a sterile sample. Daniela doesn't know how to use the equipment, so the task falls to Cheri.

We place Ana on the examination table while Cheri puts on her surgical gloves and prepares the syringe. Ana's arm is so small. She is extraordinarily thin. She cries when the needle goes in but it is all over very soon. Cheri comes through like a pro—one stick is the only penetration needed. She is nervous and dripping with sweat by the time it's over. When we're done, we lift Ana off the exam table while Cheri wipes her brow and takes off her gloves. I am standing there with a tube of HIV+ blood in my hands. It is strange to think that something so common could be so deadly.

We played with Ana for about 45 minutes before it was time for her to have lunch. We said goodbye after taking a ton of photos. I didn't think about it at the time, but that would be the last time I would see Ana until we brought her to the United States more than a year-and-a-half later.

Suzi ~ I gave your little girl a hug for you.
She sat on my lap and played patty cake.
She has big eyes, thin face, long legs.
She bites her lower lip a lot and has short dark hair ~

she needs some meat on her bones ~
We could love her, feed her and put some sparkle in her eyes.
Can you believe this?

We met a group of people from France as we walked out the back door of the orphanage. They were in Arad to deliver medical supplies and clothing for the kids. They spoke great English. We had a wonderful conversation about their organization and the compassion they have for the kids. People want to help if they have a chance to really make a difference.

We couldn't talk for long. It was too cold to stand still, and we had two vials of HIV-contaminated blood that we needed to deal with. We said good-bye to Daniela and hurried back to the apartment. Cheri and I went to our room, where we knelt beside the bed to process the blood. We waited until the red blood cells settled to the bottom before we drew the serum off the top with a syringe. We disposed of the serum and placed the remaining blood products in Eva's freezer to prepare it for travel.

The Romanian national airline, Tarom, had canceled all flights to Bucharest for the past two days because of the bad weather. Our only choice was to take the overnight train into Bucharest. It was to depart at 8:45 that night and arrive in Bucharest Thursday at 5:30 AM. It didn't sound like a fun trip, but it was our only chance to get back to Bucharest on time and still keep the blood sample frozen.

We said good-bye to Sever and Eva. They had been wonderful to us. We took a taxi to the train station, where Cheri, Dawnalyn, and I boarded the eastbound train. We called Mike to pick us up at the train station in Bucharest in the morning. He said, "Don't take the night train. It's not safe." We had no choice. We had to go. So we prepared for the train and trusted the Lord.

Sever and Eva also warned us to be careful on the train. They were worried about gypsies who might beg or steal from us. They didn't think the train was safe, especially the night train. They were also concerned that there may be no heat in the compartment. We bundled up with layers of clothing and laid out 2,100 lei each (about $3.00) for tickets in a first-class compartment.

The train pulled into the station right on time. We checked our tickets to confirm our "wagon" number and started down the line looking for the right car. We carried the luggage all the way to the end only to discover that the first-class car was on the other end of the train. We hurried back to where we started while the luggage got heavier and heavier. Just as we got to our car, the train started to pull out of the station. Dawnalyn jumped on with her bag. Cheri followed her, but the train was beginning to pick up speed. I tossed a bag up and through the open door. It hit Cheri in the back of her legs and almost took her to the floor. I ran along beside the train with my bag while Cheri cleared the entryway so I could get on. I was almost out of platform when I finally made the leap. I was on and we were safe, but everyone was exhausted from chasing the train in the cold wearing layers and layers of clothing.

We found our compartment and moved in for the night. The blood was still frozen inside an insulated bag. We hung the bag out the train window to keep the blood cold and safe until we arrived in Bucharest.

10 PM:We have been on the train to Bucharest
for an hour and a half
we are sharing a cabin with a Romanian couple ~
they seem like nice people.

Arlo Guthrie would be proud ~
The train station was something else but the train is OK.
We had to run carrying the luggage from one end of the train to the
other and then back again. Just barely caught the train. It was in the
station just 6 minutes ~ we threw the luggage on board and jumped on
~ funny now but it was no fun then ~ due in Bucharest by 5:30 AM -- I
hope we can tell where to get off. Had bread, granola and a snickers for
supper. I have only slept about 4-5 hours each night since last Friday.
It doesn't look like I will get much sleep on the train tonight.
I love you babe ~ I miss you.

Chapter Four

Homeward Bound ~ Forever Changed

Rejoice in the Lord always. I will say it again: Rejoice!
Let your gentleness be evident to all. The Lord is near.
Do not be anxious about anything, but in everything,
by prayer and petition, with thanksgiving,
present your requests to God.
And the peace of God, which transcends all understanding,
will guard your hearts and your minds in Christ Jesus.

Philippians 4:4-7 NIV

Day Six Thursday, January 21, 1993

10:30 AMArrive in Bucharest on the night train at 7:30 AM.
Mike picks us up but he waits at the wrong train.
We finally meet up and go to his mother's apartment to rest
while he goes to school and tries to get a refund
on our Tarom tickets.

I am very tired, but Ok. I call home and get Sue out of bed.
It is 11:30 Wednesday evening there
but 8:30 AM here on Thursday.

I can't stop thinking about Ana. Her eyes stay with me wherever I go. I can still feel the grip of her tiny hand around my fingers. She was the focus of our trip and now we are getting ready to return home without her. She needs so much more than she can ever receive in the orphanage setting. The staff members are afraid of her. They don't know what to do, so they let their fear rule their actions. They leave her alone to sit in her crib. I know they think, "Why bother interacting with a child who will die anyway when there are so many others?"

I move from anger to understanding and back to anger again as I think about the circumstances she faces every day. The truth is, the nurses and doctors at the orphanage are doing their best in the midst of an impossible situation. I think back to the infants we saw at Vicentiu Babes. There were at least twenty infants in a room with one person to care for them all day long. They barely got changed and fed once a shift. There is no way for them to spend time holding, tickling, and talking to any one child. It is an impossible situation.

As soon as I convince myself that the circumstances are what they are, I get angry because I am right. This is a terrible predicament for kids. I discover that my anger goes deeply into the system, the culture, the economy, and the politics that created such an inhuman situation.

Matthew 10:42
"If anyone gives a cup of cold water
to one of these little ones because he is my disciple,
I tell you the truth,
(Amen Amen) he will certainly not loose his reward."
Ana ~ skinny, no speech, no sounds, unresponsive, stiff, weak, sick
She needs a cup of cold water, quickly!!

It is not enough to just be frustrated about the number of children who are locked away from society and caught in the orphanage trap. Frustration, anger, and compassion eventually have to intersect with faith. Then there can be movement. Then there can be hope. These kids need hope.

Servanthood takes us from false Christianity
to true uncertainty
-- from safe support to risky surrender

-- from safe gods to the God whose love knows no limits.
H. Nouwen

Cheri and I had a conversation
about the possibilities for Ana ~
I think it can work if everyone is willing to make it happen.
She deserves a break and could use some love.

Day Seven Friday, January 22, 1993

I slept all night last night! It was the first time
I actually slept past 2 AM

But now I will need to adjust to the seven hour
time change in Kentucky
tomorrow will be 35 hours long.
The plan ~ up at 4:30 and gone by 5:30 to pick up Cheri.
Then to the airport by 6:00 AM for a 7:30 flight ~
one stop in Vienna ~
a layover in Frankfurt ~ then 10 hours to Atlanta ~
4 hours to clear customs, clean up a bit,
and change planes for Lexington.

I have one more week of class before I can get home to see Sue.

Just when I think there isn't much more to process, I find myself standing in front of an ancient Romanian Orthodox Church in downtown Bucharest. It is much smaller than most churches in the United States. It would be about the same size as a 75-seat chapel in most small towns across America. But this church is unique for many reasons.

It has twin round domed towers made of rough red brick. The building is built in a traditional Orthodox style. If you could look at it from the sky, the roofline would form the shape of the cross. The top portion of the cross shape houses the altar. It is rounded to create a small area where the priest does his most sacred work. It is in this altar area where the priest prepares the Eucharist and sanctifies the bread and wine so it will become the body and blood of Jesus. The Orthodox Church considers this liturgy to be so sacred; it must be done behind a screen that separates the people from the altar area.

Mike points to the towers and then makes a sweeping motion with his hand as he begins to tell about revolution in the streets of Bucharest. He remembers the night of December 23, 1989, when he was standing in front of this church. The huge open square in front of the church was packed with people. They gathered in a spirit of anger after years of frustration. They were calling for Ceausescu to show himself. They were crying out for freedom. They were finished with oppression and ready to move on into the unknown territory of self-rule, democracy, and independence. It was revolution time in the streets of Bucharest.

Mike's voice was uncertain as he told of Ceausescu dispatching soldiers to the high points surrounding the square. No one really knew what happened until it was all over. Even then, the details were often lost in the language of rumors, dreams, and propaganda.

One thing Mike knew for sure, the soldiers climbed to the bell towers of the church, where they trained their rifles on the crowd and opened fire on the people. He told of the pushing, confusion, and screaming as people ran for cover in a space where there was no place to go. The rifles were not active for long as people pressed their way into the church to confront the soldiers.

They didn't know it then, but the same thing was happening in other parts of the city. Soldiers were commanded to control the people even if it meant killing them. People were rushing the church towers to disarm the military personnel. And then the strangest of all events— the soldiers, one by one and in whole groups, gave their weapons to the people and joined the rebellion. An insane leader had ordered them to fire on the people, and now they were joining the people's revolution.

I listen to Mike's story (the people's story) unfold as we walk slowly into the church. The smell of wax, flame, and incense drifts through the air as we make our way up the steps. Mike makes the sign of the cross on his chest as we approach the outer burning candles that are placed in a large black iron replica of the church. There is an identical iron model of the church on the opposite side of the entrance. Mike explains the tradition of lighting candles as an act of prayer. One side holds candles for those who are still living. The other side represents the prayers offered on behalf of the dead.

It is silent and dark inside. An old lady with a dark scarf covering her head is on her knees in the corner. The floor looks like cold, smooth stone. She has her eyes closed and is gesturing the sign of the cross over and over again. Another older lady is standing in front of a large sacred painting. It is an icon of the Madonna Mary and the child Jesus. She touches the hand of Mary and kisses the face of the infant Jesus.

Ornate icons of saints fill the walls with the history of the early Christian faith. Every square inch of wall and ceiling space has been painted with scenes that visually teach the narrative accounts of the Bible. Everything from the gold screen to the dark, ancient, oversized wooden furnishings to the golden embroidery that accents the wall hangings has been darkened with age and with smoke from the candles that lift prayers to the heavens. Even non-believers would recognize that this is a sacred place.

It is hard to imagine anyone desecrating this place with gunshots and the activities of war. Mike silently points to one very ornate but soot-darkened icon of Saint George. It has a beautiful silver frame that is marked with the sword and dragon that have become symbolic of the life of Saint George. Mike's finger directs my eyes to the lower left side of the painting, where a bullet is still lodged in the icon. That bullet represents the price of freedom.

The revolution brought the joy of liberation along with the pain that ultimately invades every family that sends a child to war. It doesn't matter if it is an official military action, where one side anticipates the arrival of the enemy, or just a skirmish in the street. Honest liberation always comes with the sacrifice of life—from the spiritual freedom of forgiveness that comes through faith in Christ to freedom that comes in the political arena. It seems that someone always has to die.

There is a cemetery in downtown Bucharest that is dedicated to the people who were killed during the revolution in December of 1989. The grave markers are mostly large, white marble crosses. Many have photos of the person who lost their life imbedded in the stone. Each grave is decorated with flowers, burning candles, and reminders of the youth that has been lost. People go there to visit. They walk through even if they don't have a loved one buried there. They come to visit the children of war. They come to remember the price of liberation.

And they come looking for the fullness of freedom that seems to have passed them by.

I find my thoughts constantly shifting between my current experience and what I imagine Ana is doing at this moment. She was born in the year of revolution, but she has not had the opportunity to experience freedom. She was born into a nation that claims to hold the sacred in high regard. She knows only the bondage of those iron bars on her crib. She knows nothing of the basic nourishment of life that is packaged in hope and faith.

In many ways, most Romanians have experienced only a passing relationship with independence and with faith. They long for freedom but it seems to always be just out of reach around the next corner. They talk about faith but they do not know how to live with a spirit of hope.

I asked Mike why we were so easily identified as Americans were when we were walking on the street. He said, "It is because of your face." I gave him a puzzled look and he continued, "The look on your face tells us you know about freedom." That's interesting, I thought. The look on your face can be a demonstration of your personal knowledge of independence and freedom. I wonder about Ana's face. Right now is it drawn and tight? Is she biting her lower lip out of boredom and frustration? Are her eyes still held hostage by the withholding of love?

As I look at the city, I am beginning to see the depth of culture that connects everything from the process of thought to the pattern of arches inlayed in the cobblestones of the street. These are a people of rich history, deep pain, and a wealth of problems. They are willing to identify their heritage but, like all of humanity, they want desperately to hide their failings.

We walk past a small café where two men are drinking coffee in the warmth of the winter sun. Three ragged boys approach them to beg for money. The oldest of the boys can't be more than ten or eleven years old. The other two are a couple of years younger. They are all dirty to the bone, with just enough clothing to keep them from freezing in the shadows of the city. One of the men shoos them away like you would chase a fly away from your picnic lunch. The other yells at them. I don't speak Romanian but I know these were not kind words. Just then another man appears at the café door. He flies into action with a string of loud epithets directed at the boys. The boys take off on a dead run as the man from the door begins to spit on them.

Mike is uneasy with the scene. He looks at me and simply says, "Gypsies, what can you do?"

"What do you mean?"

"The boys," he replies with a shrug of his shoulders, "they are gypsies." That explains it all. They are gypsies. Because they are gypsy they deserve to be yelled at, chased away, and spat upon? I think back to one of the nurses at Ana's orphanage. I remember her warning. She not only wanted us to know that Ana was HIV+, she also wanted to warn us that she was a gypsy. She wanted us to know that Ana was not worthy of our attention because she was a gypsy. She wanted us to understand that this little, blank-eyed, frail girl might somehow bring us harm because of her genetic heritage. Ana is a gypsy. Culture is an interesting thing. It is unseen but is demonstrated in so many ways. It is under the surface, unseen by the eye, but it cannot be hidden from the heart. George Hunter says culture is the software of the mind. It is the operating system that makes the world tick. No one knows it is there, but it tries to control everything while effectively destroying some things.

We continue to walk the streets of Bucharest. It really is a beautiful city with an interesting blend of wonderful ancient architecture and communist gray concrete boxes that serve as apartments for 90% of the people. It is a city that needed to be cleaned and painted about thirty years ago.

We are running out of time so Mike offers to give us a tour of the city in his Dacia. He knows that we are interested in seeing the American Embassy where he helped Cheri and Sue with Loredana's adoption in 1991. We jump at the chance to get off our feet for a while. Romanians are accustomed to a lot more walking than we soft Americans are.

The Dacia is an interesting mechanical mix. It burns fuel that is sometimes only partly refined. It has a simple four-cylinder engine with parts that are interchangeable with any other Dacia on the road. There is no such thing as a certain year of manufacture or a specific model. If it is a Dacia, the parts all are the same as every other Dacia on the road. But it must be hard to keep them on the road, because there are lots of Dacias parked alongside the road with their hoods up and their owners under the hood.

Mike's Dacia gets around town like a buzz saw in a mill town. He keeps his foot on the accelerator even if he is at a stoplight. If it starts to slow down, he juices it a bit more to assure a good start from the light. I notice that the windshield wipers are missing. The arms are there across the windshield but the wiper blades are gone. Mike opens the glove compartment just to prove that he actually does have wiper blades. He says he keeps them there because people will steal them if he leaves them on the outside of the car. I look around at the other cars on the road (80% are Dacias); not one has wiper blades that are exposed to the elements.

This is the world of distrust that nine million Romanian people live with every day. One little girl in a tiny little white iron crib lives with it every day. I wonder if she knows what her world is really like? Maybe there is safety as well as suffering within the bars of her bed?

White

Search all you want, look throughout the sky
But you will never find "white" in a rainbow on high.
White means it's gone - the color's not there
But white keeps showing up most everywhere.

On rails that are chipped from years of abuse
To bars that are rubbed - hands with no use.
On sheets that soon will change to yellow
And the diapers - tied around that little fellow.

It seems very clean, even sterile at first,
But look a little deeper - it only gets worse.
White has a way of hiding the pain.
There is no tune for this color, no melody, no refrain.
Pretend it is clean and right in your eyes,
But white just means more cribs and more little sighs.
It only means a room filled with more of the same.
Ana's brothers and sisters ... we all go insane.

The struggle is ours - they pass it along.
They are the message - we must sing their song.
A song that paints colors high in the sky
The music of the rainbow so hope can now fly.

For the little ones...Arad 1993

We drive down the cobblestone streets toward the American Embassy. Mike is at the wheel, Cheri is in the front passenger seat, and I am shooting footage with a borrowed video camera from the backseat window. The cobblestone surface is about to destroy Mike's suspension but it doesn't seem to be a problem. He points out the front window at a large yellowish building with an iron fence around the parameter. It is the American Embassy.

I asked Mike to take it slow because I wanted to get a picture of our Embassy on videotape. I got a great shot of the front of the Embassy through the side window. I continued to shoot as Mike turned the corner so I could get the side of the building. Through the viewfinder I saw the building, the iron fence, a long line of people standing outside the fence, and some men in military uniforms with automatic weapons.

Before I knew what happened, the soldiers had stopped our car and surrounded us with weapons drawn. They were very serious and very loud. I dropped the camera to the floor while a Romanian soldier stuck his head into my open window. Meanwhile I was aware that Mike was doing some fast talking to the officer who was shouting inches from Mike's face. We were in deep trouble.

I discovered that even the most urgent prayer could not slow a racing heart. I was scared to death, and they were mad enough to make that fear come true. Mike turned quickly to look at me in the back seat. He told me that they are angry about the camera. He says it is illegal to take pictures of a military installation. Now he tells me! They wanted my camera but Mike said to just keep it on the floor. He did some faster talking, and the soldiers finally relented. I don't know what Mike told them but I'm sure it had something to do with the stupid American in the back seat who didn't know the rules. At that point it really didn't matter what he told them, I was just happy to be driving away. I didn't care if we had windshield wipers or not.

We made our escape down another narrow cobblestone street. Mike was laughing about the encounter. He thought it was fun to spar with the military. I wondered why it would be illegal for an American citizen to take a picture of the American Embassy. It turns out that there are many embassies in Bucharest. Romanian soldiers stand guard over the embassies of nations that do not have a friendly relationship with Romania. The United States and Romania are not diplomatic friends. Since there are soldiers posted at the American Embassy, it is considered a military instillation. No photos of military installations are allowed. I can guarantee they mean business.

We had had enough excitement for one day. Mike took us back to his parents' apartment, where his mother, Johanna, had been waiting for us to arrive for dinner. She had fixed a special Romanian meal for our last night together. Mike wouldn't say what it was, but he assured us it would be a wonderful surprise.

The first course is chicken soup (supa de pasare), with a quarter-inch of real chicken fat floating on top. It has wonderful noodles that are sliced very thin. Next comes a small dark sausage (cirnatsi) that is hard to cut and even harder to chew. It has a full, rich, spicy flavor that I have never tasted before. It is a specialty of the house. A family member who lives on the edge of Bucharest makes the sausage and distributes it among the clan.

Johanna is standing in the corner of the little kitchen as we sit on the stools pulled around the tiny kitchen table. She watches for indications that we are enjoying the meal she has prepared. We assure her that it is very good (foarte bun). She is all smiles. I should have known something interesting was coming up next. Johanna reached into the oven to retrieve three plates, one for me, one for Cheri, and another for Mike. They are already dished up and ready to go. She serves each person with great style and joy. It is obvious that this is show time. This is what she has been waiting for.

Johanna takes one step back as she wraps her hands in the apron that is perpetually around her waist. She wants to get a clear view so she can see every face as we take the first bite of her specialty. Mike says it is called mamaliga. The plate is filled with a huge scoop of yellow corn meal that is topped off with something white and something yellow.

One bite tells me that the corn meal is thick and very dry. A sip of Turkish coffee helps a lot! On the next bite I try the white sauce-like substance that covers the top of the mound. No question about it, it is sour cream. I mean it is real sour cream. Not processed sour cream that might top your baked potato. It is sour cream.

The yellow on top of the cream turns out to be a type of cheese. Mike thinks it may be some kind of goat cheese but he doesn't really know. He is thrilled that we are guests so he can have his favorite meal. I am just excited about the possibility of being able to eat a one-pound mound of dry corn meal mush.

Like many experiences in life, the best is yet to come. I start mixing the cornmeal with the sour cream to give it some moisture. Little by little, I am making progress while assuring Johanna that it is wonderful. Then it happens. My fork passes into the side of the mound and it starts to ooze. Johanna comes to full attention. She is excited. I had finally found it. The surprise was about to be revealed. Inside the mound of cornmeal was a hidden pocket where a whole uncooked egg had been broken and placed as a secret delight.

I was now dealing with a smiling cook, a mound of cornmeal, sour cream that had been sitting around the house for who knows how long, an unidentified melted cheese, and a raw egg spilling out onto my plate.

I could only smile as I looked up to catch Cheri's eye. She was involved in her own process of discovery. We continued to smile at Johanna and thank her for her kindness. Mike beamed as he translated for his mother, who began to glisten with joy. We were all thrilled. It took lots of syrup-sweet coffee to make it through the meal. The caffeine would keep my eyes wide open all night.

I had lots of time to think that night. The mamaliga stuck with me. I couldn't help but think about things like dysentery and salmonella. Then I began to wonder how many times Ana had eaten something that could make her sick. We knew she had never eaten solid food but we didn't fully understand what that meant.

Then it struck me. She is a survivor. She overcame hepatitis. She outlasted typhoid. She has probably eaten spoiled food. She has survived years in a tiny white iron crib, and now she will overcome the odds until we can provide a new kind of liberation. We need to liberate her body, her mind, and her spirit. But freedom always comes with a price. Will we be able to pay that price in body, mind, and spirit?

Day Eight Saturday January 23, 1993

4:30 A ~ got up early to wash up a bit.
I knew there would probably be no hot water.
Guess what ~ there is no water at all!
I think the person next to me on the plane will know
that I smell a little funny ~
I have had on the same jeans for over a week
through all kinds of situations ~
all my clothes smell like diesel exhaust and cigarette smoke.

We will leave for the airport by 5:30 after picking up Cheri at Mike's
parents' home...we make our way to the airport...
I will fly alone with my thoughts.
It promises to be a long day ~
I just want to get to Atlanta
so I can clear customs and make a phone call to Sue ~
I love you and miss you like crazy!
I thought a lot about Ana and Loredana overnight
when I couldn't sleep ~
it is a tale of two sisters.
I see Loredana in the face of young Romanian women
but I also saw her in a room of sixteen 2 year olds
being fed porridge and looking like zombies ~
they all have the same blank look that needs to be hugged away ~
such a difference between them and Loredana. ~
We are very fortunate!

The hot water situation in Romania has been a little hard for me to understand. If the community is any size at all, the hot water is all generated in a central heating plant and piped into the city through huge, insulated, above ground pipes.There is no home hot water heater or family furnace downstairs in the basement. The hot water is heated at the plant and distributed to every home, apartment complex, school, public building, and business. The hot water is available for washing dishes and taking a bath. But it is also the only source of heat in the winter.

In Bucharest, the capital city of over 2,000,000 people, the hot water is turned off during the night. In fact, most of the city has hot water only from about 6:00 to 9:00 in the morning and again from

6:00 to 9:00 in the evening. Somewhere on the edge of the city there is a huge faucet with a valve that shuts down the whole system.

The result is a very large population that organizes their lives around the availability of water. I don't know who thought up this idea or how long it has been going on, but it is a powerful way to control the people. It is like a population-size mind game. Will we have water today? Will it be hot? Will there be heat in our building tonight? No one has the answer and no one has control. There are no thermostats on the living room wall to adjust the temperature to comfort. It is either on or off. This time it is off. There is no water in the city, hot or cold, at 5:00 in the morning when I need to wash up for the 24-hour plane trip home.

The concrete floor of the apartment is too cold to walk on. The temperature inside must be somewhere in the forties or low fifties. It is too cold to spend much time worrying about the politics of hot water. This is a time to get dressed, and fast. No wonder Ana was always dressed in so many layers. I wonder how many nights she has slept in the cold? How do you avoid sleeping against those cold white iron bars? How often does she get a bath? Is the water ever warm? And who would she complain to if she wasn't being cared for?

She has learned the fine art of shutting down. There is no sense in crying if no one answers. There is no incentive to interact if there is no one to smile back at you. There is no reason to rely on anyone outside your own personal environment. She has learned to shut down. She has learned to survive. I don't want to ask the obvious question. How long can she hold on?

I am going to be traveling home alone today. Cheri will be staying a few days longer to take care of some details with the American Embassy in Bucharest. She will be carrying the insulated container with Ana's frozen blood sample to Atlanta. The insulated bag looks like an elementary school lunch box but we all know it contains life-and-death information.

Otopeni Airport looked the same. Check-in for my flight home went well, but I did have an interesting experience in the terminal bathroom. It seemed to be a unisex arrangement, with the convenience of clear glass doors that leave everyone and everything exposed to the whole terminal population. The floor was sticky. I really didn't want to know why it was sticky. I just wanted to get in and get out. The lady in the stall next to me is either cleaning or searching for the toilet paper that does not exist. I guess if Ana can spend years sitting on a 2×12 wooden plank with eight holes, I can live through this. Who would ever guess that Ana would bring me encouragement in a public bathroom setting?

The plane was full. My seat assignment, 22 D, put me next to a very interesting young man. He works with a mission agency known as *Jesus The Hope of Romania.* He had been in the country for the last two weeks providing administrative support for their mission projects. We discovered that we both had similar concerns.

Neither one of us had had a bath in the last few days, and we were both hungry for something American. We shared my last granola bar and talked about the children of Romania. We exchanged book suggestions and worked on a Sunday School lesson that he was going to present in his home church near Cincinnati. We talked so long that we discovered we had missed the announcement that our plane was being diverted from Vienna to Munich due to high winds in Vienna.

11:00 AM Bucharest time 1-23-93 Still on the ground in Munich ~
I am going to miss my connection in Frankfurt
if we stay on the ground much longer ~

As I tell the story of our experience in Arad,
it becomes more and more clear about the intervention of God
and the grace we received ~ it is not just the grace
of being in the right place
at the right time (that makes God work by accident) ~
It is not an accident, it is the work of prevenient grace ~
God was already there when we got there!
God had already intervened by preparing the way!
Best yet ~ God is still there!

1:30 Frankfurt time 2:30 PM Bucharest time
About to get going on a Delta flight to Cincinnati.
Missed my connection because of the delay in Munich.
I could have jumped to the plane
as it pulled away from the terminal.

It was that close.

1½ hours in line to rebook to Cincinnati ~
then the computer goes down just as I get to the counter.
Try to grin and bear it but one "ugly American" began demanding
special treatment. I apologized for him to the agent he mistreated.
Then guess who I am seated next to on this flight ~
Mr. Ugly American himself.

I plan to sleep lots! Hope my bag makes the same flight.

Chapter Five

The Joy of Spiritual Discovery

Do not conform any longer to the pattern of this world,
but be transformed by the renewing of your mind. Then
you will be able to test and approve what God's will
is --- his good, pleasing and perfect will.

Romans 12:2NIV

Day Nine Sunday January 24, 1993

10:00 AM Wilmore, KY My flight got into Cincinnati
on time at 5:15 PM
then a quick run through customs and I catch a 6:00 connection to
Lexington ~ got a chance to call Sue to give her a quick update on the
flight rescheduling.
Brian ~ a student at the seminary ~
picked me up at Blue Grass Field.
He recognized my hunger ~ we got a hamburger at the McDonalds drive
through ~ then to Beeson Manor to call home.

Took a long shower and slept from 9:00 to 6:00 AM ~
up early to eat oatmeal and drink American coffee.
Will attend Wilmore United Methodist Church at 10:45 ~
meanwhile I catch up on my reading.

Eddie Fox ~ Let The Redeemed of the Lord Say So!

Dr. Charles Bertrand was preaching at the Wilmore United Methodist Church in Wilmore, Kentucky that morning. His text came from Hebrews 12:15 with support from Matthew 4:12-23. His theme was appropriate for me and my Romanian experience. He spoke about listening for and hearing the voice of God.

I couldn't get the sounds of the children out of my ears. The taste of their struggle was lingering in my mouth. The sights were still flashing before my eyes as I thought about Ana and her abandoned roommates. I wanted to hear from God to clear up my confusion and to equalize my emotions.

Rev. Bertrand said there are times in our lives when we experience spiritual deafness. That wasn't my immediate problem. I was encountering spiritual and sensory overload, but I was still listening for the voice of God. Then he asked a profound question, "Is the voice of God as crisp and clear as it has been in the past?" He drove the question home by closing with an invitation to repeat this phrase, "I will hear what God the Lord will speak."

I made that commitment to hear what God would speak, and I will never regret that resolution.

Everything became clear at that instant. It was a God moment. The 2,000-seat Wilmore United Methodist sanctuary surrounded me like the arms of God. My pew was no longer a hard blonde board. My thoughts were clear. My emotions were on the surface but they were all flowing in the same direction. Finally, the message was clear.

I had been waiting for the voice of God to give me direction. I kept asking, "What's next, what shall I do? But all along I had been listening for the wrong voice. I don't know quite what I had been expecting, but now it was clear. I needed to listen for a voice I had never heard.

It was the voice of a child who could not speak. It was Ana's silent voice that had already spoken God's word to me. She was saying, "Come and get me. Don't leave me. Be the hands and heart of God before it is too late." Now it was my turn to listen to her silence. It was my turn to give ear to her silent cry. I didn't want to miss a single word from those silent lips. Her words changed my life forever.

In my moment of discovery, the 1966 lyrics of a musical poet named Paul Simon flashed through my mind.

Hello darkness my old friend,
I've come to talk with you again.
Because a vision softly creeping
Left its seed while I was sleeping,
And the vision that was planted in my brain
Still remains
Within the sounds of silence.

I explored the lower level of the Asbury Seminary library to discover the history that surrounded my experience. It is interesting that we walked into a nation that was closed to western passports. Even Newsweek, January 1, 1990 doesn't really know what happened in the Christmas revolution of '89. The world was closed, and then the wall came down and there was hunger—hunger for the Spirit. I can help free those who are hungry.

Day Ten Monday January 25, 1993

My Doctor of Ministry class at Asbury Seminary started first thing Monday morning. The class was planned for off-campus students who would do advance work before attending class and then follow up the class time with projects, reading, and papers. The daily class sessions were scheduled for at least six hours. I was looking forward to this particular class. The Seminary brought in a guest lecturer for this special session. Dr. Eddie Fox was going to be the instructor for a class titled *The Congregation As Evangelist.* Dr. Fox was well known around the world as an evangelist who specialized in church planting and church growth through the World Methodist Council and the United Methodist Board of Discipleship.

I have admired Dr. Fox and his work for many years. His powerful faith and positive spirit is always evident in his preaching and in his writing. I remember hearing him preach at a conference in Breckenridge, Colorado in 1988. Three years later, Sue and I heard him preach a powerful message at the 1991 World Methodist Conference in Singapore. But now it had been almost five years since I first crossed paths with Dr. Fox and here he was teaching a class at Asbury while I was doing my best to recover from my first experience in Romania.

The conference Dr. Fox preached in Breckenridge, Colorado in 1988 had a life changing impact on my ministry. I remember the day because it was a turning point in my life. I had been ordained for just a couple of years when I became disillusioned with the whole system. I was ready to resign my credentials and become an active layperson in the church again. Then Eddie preached. I don't remember all the details of his message, but I vividly recall his challenge to stand firm and stay strong in the faith so others would have a positive witness of Christ. The challenge moved my heart.

I had a file with all my ordination credentials with me at that conference. After Eddie's message that afternoon, I put the file back in my brief case and took them home to hang them back on the wall of my office. Thanks to Eddie, they have been there since that day.

The Classroom

The class group was small. There were only eight of us in the whole group. We were all seated together in a horseshoe arrangement with Eddie at the opening as we shared our names and spiritual journeys at the opening session. I told the group my name and the church where I served as pastor. I mentioned briefly that I was a bit tired because I had just returned from a quick trip to Eastern Europe.

No one seemed overly interested in the trip so I let it go at that. But the truth is, I couldn't let it go. My experience in Romania had a grip on me like Hulk Hogan with a hammerlock on the Macho Man. But this experience wasn't just scripted to appear genuine. This was the real thing. It was a spiritual adventure that would last a lifetime.

The informal group introductions were finished, but I must confess I didn't hear a single name. I couldn't move beyond my own thoughts. I was reliving the experience of walking into Ana's room. The image of her ragged hair and the smell of her surroundings would not leave my thoughts. I was a hostage in my own contemplation. Dr. Fox was outlining the journey he had planned for the class over the next five days. But I was still locked in my own silent world as I weighed the speechless world where Ana was held captive.

Most people would say their biggest fear is speaking in a public setting. I overcame that fear long ago, but I still had one lasting fear that was becoming reality right there in Dr. Fox's classroom. It was the fear of weeping uncontrollably in public.

There I was, caught in my own thoughts and overwhelmed with compassion. I was wondering about Arad and thinking about what Ana was experiencing at that very moment. Suddenly I realized that I was weeping. The tears would not stop. They fell from my cheeks onto my open notebook.

There I was, sitting in Eddie Fox's class, and I felt an overwhelming desire to serve the people of Romania. Reaching out is not my idea—it is God's idea. Evangelism is God's idea, church is God's idea, and ministry is God's idea!

At the lunch break I took a moment to speak directly with Dr. Fox. He was warm, attentive, and extremely helpful. I gave him a nutshell version of my experience and the child I left behind in Romania. He offered his help and told me about his work of starting churches in Bulgaria and Russia through the World Methodist Council. His heart resonated with my situation and I felt a deep peace. I will be forever thankful for his words of compassion and encouragement.

Four years later I arrived back in Wilmore for my Doctor of Ministry graduation exercises. I was filled with joy when I discovered that Dr. Eddie Fox would be receiving an honorary Doctor of Ministry degree in recognition of his work in the area of world evangelism. Dr. Maxie Dunnam, President of Asbury Theological Seminary, presented the hood and the documents to Eddie and then invited the "country preacher" to address the graduating class.

I remember his warmth and his words of encouragement. He spoke about how God had called him from a simple life to the life of a world evangelist. He said God could touch people wherever they were. He said God spoke to him as a country boy who "lived so far out in the country we had to go toward town to hunt." I listened to every word.

When my name was called to step on the platform to receive my diploma, I went straight to Eddie Fox to shake his hand and say thank you before I received my diploma from Dr. Dunnam.

Eddie Fox doesn't know that he became Christ to me twice in my life. Once in Breckenridge when I was discouraged and ready to give up, and then again when he listened to my tears and heard the sounds that came from Ana's silence. Thank you, Eddie.

God has such a wonderful way of bringing transformation into our lives. Sometimes we are seeking the freshness of a new experience, and other times we are blessed to discover something we weren't even seeking. I rediscovered the truth of the prophet Isaiah. "The wolf will lie down with the lamb, the leopard will lie down with the goat, the calf and the lion and the yearling together; and a little child will lead them."

The child Jesus ultimately came to bring a radical peace, a spiritual peace, and a transforming peace. Now Jesus was using another child to provide transformation in my life. It was one of the most renewing experiences of my spiritual journey. God was around every corner, and there were lots of corners to explore.

Steve Beard of Good News Magazine gives Ana a much needed hug

Cheri Walker, Mr. and Mrs. Cionca (Ana's birth parents) and
Randy Jessen outside the state orphanage in Arad, Romania

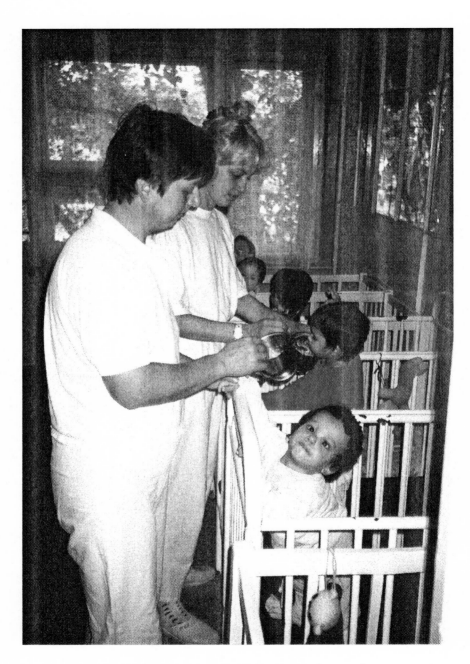

Lunch time at the Dystrophic Hospital in Arad

The horse drawn food delivery system at the back door of Ana's orphanage

Ana's first day (September 8, 1984) outside the orphanage with Randy and Cheri

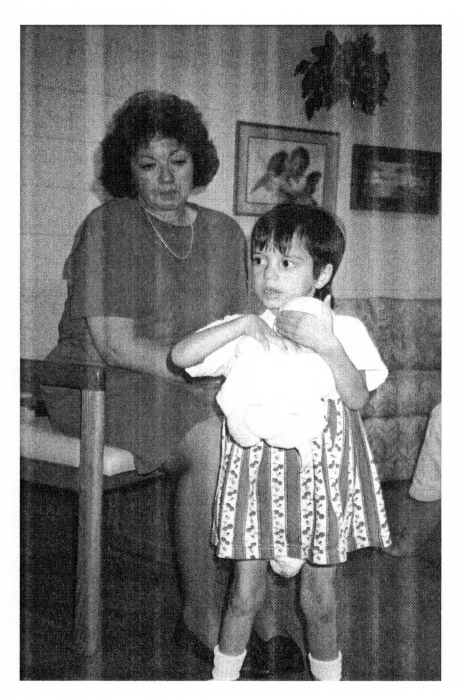

Ana's wobbly legs arrive at the Broomfield United Methodist Church

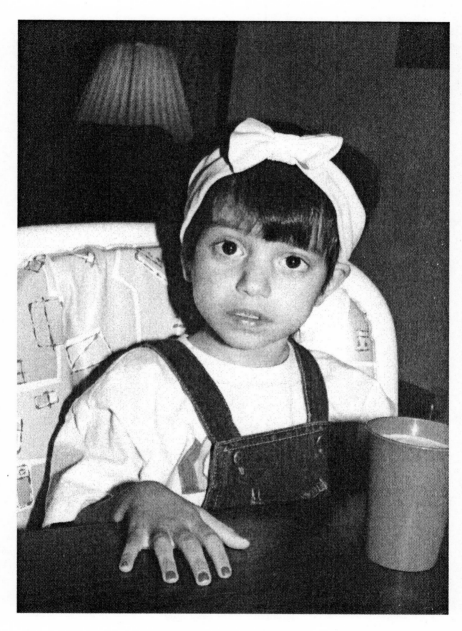

Blank eyes and freshly polished nails. AnaBabe...I love you!

Ana's first Christmas in Colorado with her signature pink high top shoes

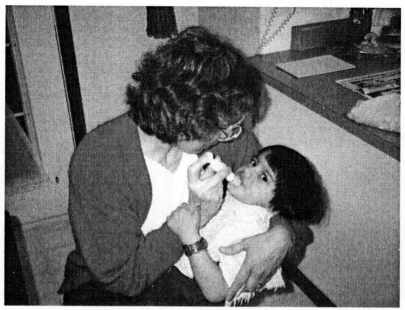

*Sue personally fed Ana by hand five to six times each day
for the first several months, one syringe of formula at a time.
Nutrition and a mother's love make all the difference.*

*The kids at Vicentiu Babes orphanage have another yummy lunch while
Loredana Walker watches from the end of the table with great interest.*

My first preaching experience in Romania (September 4, 1994) at Speranta Baptist Church with Pastor Romica Iuga. It is the pulpit where Billy Graham preached in 1985. Over the years, Romi and I have enjoyed preaching together in many Romanian congregations.

October 2005, Randy, Ana Sue, and Sue Jessen. Ana is sweet 16!

Chapter Six

Family Matters

God sets the lonely in families,
he leads forth the prisoners with singing.

Psalm 68:6a NIV

Nineteen Months Later

A remarkable thing happened on Thursday August 18, 1994. I heard the phone ring but I knew Sue would pick it up. Then she called out that the phone was for me. I picked up the portable phone in the living room and discovered Cheri on the line. She said seven words that marked the start of another powerful adventure of faith. She said, "Are you ready to go to Romania?"

"Sure, what's up?"

"The American Embassy just called. They are going to issue Ana a visa."

I couldn't believe my ears. After all this time, after so many delays and denials, they had now consented to issue a visa for Ana to come to the United States in spite of her HIV status. "When can you be ready?" I heard the words in the phone but it really hadn't sunk in yet.

I responded, "When can we get a flight?" One week later we were on a United Airlines flight to Bucharest to bring Ana home to Colorado.

Ana had been a part of our family from the very first moment. We kept in touch with her condition through friends. But our primary contact was through our daughter Cheri. She had committed herself to bringing Ana to the United States. Cheri made four additional trips to Arad with various members of our family just to see Ana and to move the process off dead center. She gathered support, knocked on doors, talked to Senators, and decided that she would never give up until Ana knew that someone loved her. She filed visa applications that were turned down three times because Ana was HIV+. Cheri and her determination kept our family believing that one day Ana would finally be home.

> *Thursday 8-25-947:15 A Plane takes off for New York.*
> *I say goodbye to Suzi and move forward with the*
> *dream of bringing Ana home ~*
> *it has been her home forever and we have been her family forever ~*
> *she just doesn't know that yet. We pray that we would be washed clean*
> *of any selfish gain and that her needs would come first ~ I pray that she*
> *will be able to overcome the trauma that is about to happen to her.*
> *Steve Beard will meet us on Saturday ~*
> *we will arrive about 24 hours earlier.*
> *Thank you Lord for bringing us this far.*

We are on our way to make a dream come true for our family, but we are acutely concerned about what Ana will experience. She has lived her whole life in the orphanage system. It is the only home she knows. The things we see as depressing and painful make up the only reality she has ever experienced. She must think that her world is normal. Actually, we don't know what she thinks. We don't know what goes on behind those beautiful dark eyes that sometimes look so vacant and lost. We don't want this adventure to be any more traumatic than it has to be. But we know that the events we dream of are destined to change everything she knows and understands about life.

I think back to the year of her birth. 1989 brought revolution and liberation to the streets of Romania. I pray that this trip will result in emancipation of mind, spirit, and body for a little girl who speaks only silence and knows only confinement. She has been a part of my daily prayer for the last twenty months. She has been a part of our family since the moment we heard her name. Jesus said that God sent Him "to proclaim freedom for the prisoners and recovery of sight for the blind, to release the oppressed, to proclaim the year of the Lord's favor." Surely this is the year of the Lord's favor for Ana.

Our landing in Bucharest revealed a different atmosphere from my previous trip. There were still lots of armed military present, but they were not nearly as visible. The airport was a bit brighter and the floors were a bit cleaner. It was still loud and confusing, with lots of jousting at the baggage area. Steve Beard, of *Good News* magazine, would be joining us soon. His flight was due about twenty hours after our arrival.

Steve and I would be staying with Dr. Joseph Bati and his wife Helen while we were in Bucharest. Dr. Bati is a Ph.D. professor of economics at the University of Bucharest. He also served for a time in the Ceausescu government as an under-secretary of economics. He was trained at Karl Marx University in the former Soviet Union.

Dr. Bati is a wealth of information. His unique area of expertise is in the study of labor and labor management. His writings are published often in economic journals, and he is a regular commentator for newspapers and television reports. He is a man of intellect and dignity who actually retired from active teaching several years ago, but the inflation rate devoured his university pension. His pension was almost luxurious under the communist income and price controls of Ceausescu's government. But now inflation is running over 10% per month and he is back to work so he can pay his rent and care for his family.

The average salary in 1993 was $36.00 per month. Now, less than two years later, the average salary is almost $50.00 per month, but inflation and high prices have consumed far more than the outward increase.

Dr. Bati likes to talk politics but he doesn't like to talk about the Romanian Revolution of 1989. In fact, he never refers to the revolution at all. Instead he talks about it in couched terms. He prefers the language "the events of December 1989" instead of the revolution of `89. The distinction is important to him. What happened in 1989 was an "event" not a "revolution." He believes it will take a full generation for the economy and the political climate to stabilize.

Steve Beard is a political thinker, so he and Joseph Bati hit it off. Steve can't believe that he, a conservative Republican, can be drinking coffee and exchanging ideas with a graduate of Karl Marx University on a seventh-floor balcony in Bucharest. They discuss everything from American investments in Eastern Europe to international foreign policy strategies

On Monday morning, August 29, 1994, we were at the American Embassy shortly after the consulate offices opened. Our friend Dan Nestor took a day away from his work to drive us and to serve as a translator. They don't come any better or more faithful than Dan. He took us to the side door and gave us instructions, but he couldn't go with us because his Romanian passport would not allow him entry. Without an American passport, people waited in a block-long line just to enter the building to pick up visa application forms. By the time we entered the building, the line extended down the block and around the corner. Armed guards watched every move of every person in the line. Mostly they did a lot of pushing and shouting at people to stand in a straight line. The process didn't make sense, but our American passports provided an instant entry into the facility.

I remembered this side of the building from my previous experience here. This was the place where the Romanian military stopped our car and wanted to confiscate my video camera because I was taking pictures of the Embassy building. I looked closely at the Romanian soldiers who were stationed at the gate. I wondered if any of them were on duty that day almost two years ago.

My heart was swiftly reminding me of my excitement as we entered the Embassy. I think it was the result of the Romanian soldiers coupled with the thrill of finally obtaining Ana's ticket to freedom. A visa is a powerful tool. It can come in the form of a simple stamp or a formal document designed to discourage forgery. It can hold you captive or allow you the ability to cross borders. It can keep your mind oppressed or give you the confidence of independence. We were about to discover that nations of all sizes play political visa games to reinforce their sense of sovereign control.

Steve, Cheri, and I cleared the metal detector and received permission to enter the main Embassy operations area after stating our business at a little glass booth. I felt a little guilty that we got in so easily while so many Romanian citizens were standing in a guarded line. But after all, this is the American Embassy, I am an American citizen, and this is a place of refuge for American travelers. Then we came to the counter where the visa was to be issued and began to learn about the reality of small people in powerful positions.

The three of us approached the counter like happy musketeers who were about to get their reward for being diligent and persistent. Cheri laid out her paperwork and told the clerk what we needed. The clerk immediately went to get her supervisor who came to take care of the situation. She was very familiar with the application for Ana's visa. She was the one who had denied the application in the past and she was not thrilled to know that State Department officials had gone over her head to authorize the visa for this little girl.

We were told in no uncertain terms that this was not her choice but that she was being forced to issue the visa. Since Cheri was the primary applicant, the supervisor insisted that she would only do business with Cheri at the counter. Steve and I would have to leave. We consented and went to a seating area just around the corner.

The supervisor was not satisfied with our location. She left the counter, came outside her work domain, entered the seating area, and restated her request. She would only deal with Cheri and we would have to leave. We politely told her that we thought we had left the area and that we would not be in the way or interfere with the process. She was angry. She insisted that she would not issue the visa until we were out of the area. "What does that mean?" we asked. She replied that she would not issue Ana's visa to enter the United States until Steve and I were out of the building.

She knew that everything depended on the visa. We knew that she could make things complicated, so we left the building. We went out past the Marine Corp guards, past the Romanian military, and stood across the street facing the long line of Romanian citizens who were trying to get in. We had been thrown out of the American Embassy. So much for a safe haven. So much for American territory on foreign soil. We were aliens outside the gate.

The good news is we got the visa. Ana was free to travel to the United States on a renewable visa that allowed her to stay six months at a time for up to two years. At this point, her life expectancy was anywhere from six months to one year. We had plenty of time to take her home and love her.

They discreetly marked "HIV+" in the lower corner on the last page of her passport as a reminder of the things that were really important in life. Officially it was still illegal for people who were HIV+ to enter the United States. The original law was written in the late 1980s in response to the migration of Haitian refugees who were HIV+. Congress passed a law prohibiting any entrance for people holding a foreign passport in an attempt to stop the drain on the American medical system.

We don't know for sure, but it is possible that one event in particular may have loosened this law just a bit. In the summer of 1994, just six weeks before we received word that Ana's visa would be issued, the United States hosted the Gay Olympic Games. Hundreds of gay athletes from around the world were allowed entry into the country to participate in the games. Apparently, some of those athletes were HIV+, but the State Department chose not to enforce the HIV+ no-entry rule.

Our next task was to get a German transit visa for Ana. Any Romanian who even changes planes in Germany must have a transit visa. Our return flight schedule called for us to spend part of one night in Frankfurt. We took a taxi to the German Embassy in Bucharest to find another long line outside the building. Once again, Romanian citizens were lined up to apply for visas. Life in this part of the world seemed to revolve around visas and official documents.

We got in line knowing that it might take most of the day just to get inside to obtain the application. A Romanian man in line with us noticed that we had blue American passports instead of the traditional Romanian maroon passport. He spoke enough English to tell us that we were waiting in the wrong line. He directed us with sign language to check in with the guard who was regulating the flow at the front of the line. The guard took a quick look at our passports and invited us to enter the embassy. He even smiled as he moved us through in front of the rest of the line.

Inside, the visa application was fairly simple except that we needed to pay in German Marks, of which we had none. Steve and I got a taxi and found a bank that handled money exchanges. The line at the exchange window was fairly short. We took our place only to discover that everyone else was moving right to the front of the line. They pushed ahead and crowded around the small opening at the teller window. Apparently Romanian culture does not consider cutting in line to be a huge social miscue. We quickly learned how to play the game. We converted $20.00 to Marks and made our way back to pay the five-Mark visa fee at the German Embassy. We had to return on Friday to pick up the actual visa. The lady at the counter was helpful and sympathetic. She couldn't have been more customer friendly.

It was an interesting day. Our American stereotypes presuppose that German people are harsh and maybe a bit rude. We assume the American "safe haven" Embassy would be our best friend in a time of need. It turns out that both assumptions, like many conjectures, are false. The Americans threw us out and the Germans welcomed us with open arms.

Cheri caught a plane to Arad on Tuesday so she could make some contacts and pave the way for Ana's release. Steve and I had more time to walk the streets of Bucharest and talk with Dr. Bati. We were to hop a train for Arad first thing Wednesday morning and meet Cheri there in the afternoon. We discovered a little outdoor coffee place on the corner of Dorobantilor Street and Dacia Boulevard by the Piata Romana Metro station. We had plenty of time to talk and process the experience.Steve and I will catch the 7:00 AM train from Bucharest to Arad. Our mission was just beginning.

Chapter Seven

Silent Child

For you created my inmost being:
you knit me together in my mother's womb.
I praise you because I am fearfully and wonderfully made;
your works are wonderful, I know that full well.

Psalm 139:13-14

Wednesday August 31, 1994 on the "express" train from
Bucharest to Arad with Steve Beard.
Cheri flew to Arad on Tarom to get a head start
on the process for Ana.
Steve and I will arrive this afternoon.
I was watching very poor farmers from the window of the train.
Broken down horse-drawn wagons loaded with grass hay
that has been cut with a scythe and turned to
dry with a wooden pitchfork.
Husband, wife, children ~ the entire community
digging potatoes by hand.
Faithful people of the soil planting an 8X10 plot of cabbage
on the hillside of the train tracks.
Workers taking a break in the shade of a ruined
concrete block building that no longer has a roof.
Rust is everywhere.

I had an urge to capture their strife on film.
I wanted to take it home to show others.
I had the desire to say "see how poor
and backward the people are" ~
then I realized that the camera I held in my hand ~
a small, simple, point and shoot camera ~ was worth a month and a half
wages for the average city worker. Perhaps it is as much value as six
months pay for these farmers ~ they live a simple life.
I took no more pictures.

A young couple got off the train. It was a major stop.
A community with a factory, houses, a "Banccoop"(whatever that is),
old men on bicycles,
children playing in water, people talking on the street,
and a haze of poverty everywhere.
They crossed eight sets of tracks to get to the street.
Hand in hand they helped one another
over the rails and through the ties.
They climbed over a rail of steel pipe that had been
painted until it had more blue paint than steel pipe.
A white rail leads them to a small unmarked street.
They have been there before.
No one looks for direction or orientation.
They just walk. Slowly, hand in hand through the street.
Sidewalks are present but they walk in the street.
Up the small hill past the Banccoop where they veer to the left and
finally out of sight behind an unknown building that desperately needs
paint. They are gone.
A horn honks with anger, a man with a bag in both hands
hurries across the street.
Today is not a good day to get injured in the street.
But it is a good day to walk hand in hand until you are
slowly out of sight.

The train station in Arad was just as old, dark and dirty as the last time I was there. Actually, it was the winter of 1991 when I last caught a train at this station. I remember the cold, the dirt and the masses of people. Today it is warm but the air still has that oily smell and there are more people in the station for the early afternoon rush.

I was thankful to see Cheri and our translator, Mr. Black, waiting for us at the station.

Sever and Eva Jurca will host us in their apartment. In 1993 we all shared the same apartment. Now they have moved to a "villa" by the Mares River just past the Boulevard. I am thankful that they still own the old apartment. Steve, Cheri and I will enjoy the expanded space as we share the Jurca apartment. We will introduce Steve to Ana in the morning. I went to sleep wondering if Ana remembers us. I wonder if she has changed in the last year and a half since we saw her last. Does she know we are coming?

I woke Thursday morning with a bit of anxiety. We were about to take a huge emotional leap of faith. Our family had decided that Ana would live at our house so she would be closer to the medical care she needed. Sue and I would be her primary care givers. We would be in charge of being with her daily as we prepared for her to know that she was loved before she died. We had been "empty nest" parents for several years. What would we do with a 5½ year old? And how would people around us handle the HIV factor? The future was uncertain but the mission for today was clear. I wanted more than anything to see her once again.

We were at the back door of the orphanage just after the change of shift at 9:00. The windows were open and the sounds of children spilled out into the back yard. A lady in white answered the door and invited us in with a smile. She recognized Cheri from her previous trips to see Ana. Cheri had already made four additional trips just to see Ana and to provide specific care for her.

The attendant in white directed us upstairs. She pointed to the visiting area where we would wait while she went to get Ana. Every adult in the building knew why we were there that day. We were there to finalize Ana's release from the orphanage system. We were there to surround her life with the blessings of freedom and love. Every child in this convoluted system deserves the same gift.

A different worker in white brought Ana to the visiting room. Ana has grown. She is taller. Her hair is cropped short. She is wearing overalls and tiny green sandals. She is walking a bit better but she still moves like a tiny pregnant lady. Her hips are pressed forward, arms extended for balance, and her feet move in a waddling shuffle. Her lower lip is almost raw from biting it with her upper teeth. She makes occasional eye contact and even offers a hint of a smile when Cheri tickles her tummy. The attendant waits for a few moments and then disappears down the hall.

We have a chance to play with her and to attempt to access her health. Her eyebrows are knotted in a pained expression. It looks like a mixture of confusion and exhaustion. I pick her up under her arms and sense the fragile nature of her body. Her layers of clothing and heavy diapers hide the fact that she weighs much less than a typical 5-1/2 year old. Later we would sadly learn that she weighs only eleven kilos (about 24 pounds).

She makes a funny little squeaky sound with her lips when she is excited. Occasionally she will giggle. Otherwise there is no sound at all. She is the same silent child we saw over a year ago. She has no vocabulary, no language, and not even the sounds that infants make in playful response. A little giggle and a tiny squeak are the only sounds of her life. The staff has never heard her make any other noise. She doesn't cry, she doesn't laugh, and she has no words of her own. Yes, we will be able to love her. Yes, we will be able to care for her physical needs. But will we be able to give her a voice so she can experience the joy of communication? Will she ever have a voice so she can teach others how to love?

We cannot stay long because we must meet with the director to establish a plan to remove Ana from the orphanage. We display her Romanian passport and her freshly issued American visa. We have documents signed by the birth parents, a social services document, and letters of support to demonstrate our intent. We have one more hurdle. We must present our case before the Arad County Commission on Child Protection. They meet once each week on Wednesdays. We will have to make fresh travel plans, stay a few extra days, and take a chance that the Commission will approve our plan.

Meanwhile we consult with Sever, our attorney. He gives us a plan that might work. It is complicated but he thinks we have a chance. We will go to visit the birth parents. We will meet with Ana and Ovidiu Cionca at their home in Curtici. We will ask them to appear with us at the Commission hearing to request that Ana be released to their custody as the birth parents. The orphanage will release

Ana to the natural parents and they in turn will sign Ana over to our care so we can take her to the United States. Sever is concerned that the Commission will see through the plan and refuse to release Ana because of her irredeemable HIV status.

We talk over the plan while sharing a quick lunch of Romanian bread, Coke, "imported" canned tuna, and granola bars. In addition to the hearing with the Commission, we must return to Bucharest overnight to pick up the Ana's German transit visa so she can land and change planes in Frankfurt.

First, we need to hurry back to the orphanage so we can see Ana before we head back to Bucharest. The staff has been waiting our arrival. They have prepared Ana for the visit with a bath, freshly combed hair and she is dressed in a cute little dress that has red and white vertical stripes. They are waiting for us outside under a big tree that provides shade to what once was a playground. Ana's eyes are squinted from the sun. The constant dazed and confused look is still present in spite of the fresh air. She is unsteady on her feet with the dirt terrain of the former playground. And she seems uncertain of why everything is different in her life all of a sudden. She is out of her crib, walking outside, meeting new people, and getting dressed for the occasion. We wonder out loud if she has any insight into the events that are about to unfold.

We ask for permission to take a few photos. The attendant gives an affirmative nod while Steve produces his camera and shoots a couple rolls of black and white film. He captures her puzzled expression and her exposed knees that look like large knots on a spindly tree branch. It is our first unhindered view of her legs. No wonder she walks with such a strained gait. Her legs have no muscle at all. She is bone thin from ankle to hip. Her knees have a silhouette that erupts from the thin line that is her leg. To top it off, her knees don't seem to bend. They are stiff from years of being held hostage behind those tiny white iron bars.

I took hold of her hand to lead her down the shattered sidewalk. She wrapped her tiny fingers around my hand and came with me. She didn't change expression or look up to see who was accompanying her on this journey. She just obediently took the hand that was offered and submitted to the ritual. Her fingers are so thin and transparent. They almost look artistic. But there is nothing poetic about a hand that has only played with the dust in the sunlight and never held a toy.

The attendants let us know that it is time to go. We must leave and Ana must return to her room to be with the other children. The truth is, she will go back to her room but she will not interact with the other children. She will be alone with her own thoughts. Her fingers will fly across her path of vision in an intricate fluttering motion. Her tiny transparent fingers will furiously fan back and forth across her face to break up the light and stimulate her brain. She will spend the rest of the day amusing herself watching the light shift between her fluttering fingers as she drifts into the darkness and finally finds relief in her dreams.

I wonder what she is thinking? I wonder how she will sleep tonight? Does she know we will be back? Does she know we are about to remove her from every thing she has ever known? Will she understand? Does she know that we love her?

My mind is racing as we regroup at the apartment. We need to extend our airline tickets so we can be present for the Wednesday hearing with the Commission. We need a plan for the hearing. And we need to phone home so the rest of the family can pray for our success.

I am making a list of the things we must do when the telephone rang. It was my friend Ioan Redis. We have corresponded by mail several times since the day we first met in the cold sanctuary with his high school choir students. Our letters have established a long distance friendship that was sparked at our first meeting. He wants to pick me up so he can show me the Christian High School and introduce me to his family.

Our first stop on a whirlwind tour of the city was historic. We were at the first Protestant church built in Romania. The Baptist congregation is planning to tear down the original building as it completes a huge building project. We make our way through the construction site to the back of the building. The Christian High School and the fledgling Christian Nursing School are both housed in this facility. Ioan, or Nelu, (as he is known among friends) is the Director of both institutions. They have three rooms and a small office ready for class groups. They have two additional classrooms, a library and a set of bathrooms left to finish.

Both schools are private and both schools are the first of their kind in the country. They are unique in a country that is still operated by a Communist government and organized around the principles of socialism. As a private school the students pay tuition of 4,000 lei (about $4.50) per month. He is desperate for scholarships so his students can remain in school.

During our tour, Nelu mentions that his pastor has invited me to come to their church this Sunday to share with the congregation. It is the Speranta (Hope) Baptist Church where I first met Nelu and distributed Bibles to his students. It seems unusual that the pastor would invite someone he had never met so I try to casually direct the conversation away from the invitation. That tactic doesn't work. Nelu insists that I have been invited to preach and that he will introduce me to the pastor on Sunday morning. It will be a communion service so I have been asked to say a few words about the Lord's Supper. I agree and receive special dispensation to dress casual since I didn't pack a coat and tie.

He drops me off at our apartment after sharing a prayer with me in his car. He prays partly in English and partly in Romanian. He asks boldly for safe travel, a successful mission, and he adds a blessing for Ana.

He doesn't know it now, but he will meet her next week. Right now his face reveals the questions of his heart. Why would people come all the way from the United States to rescue an irredeemable gypsy girl who has AIDS?

Why? It is the eternal question that links doubt to faith. Our answers always come to life from the dust of our doubts. Why did Jesus say, "Let the children come to me?" Why did he give his life on the cross? Was it really for me? Or was it for you? Or was it for Ana?

Chapter Eight

Escape to Home

"As a deer pants for streams of water,
so my soul pants for you, O God."
Psalm 42:1

It is Friday September 2nd on the confusing streets of Bucharest. The sun is warm but even bright sunlight cannot warm this city that has been the center of Communist control for over a generation. We are in a hurry to get to the German embassy before it closes at noon. We are almost running along the sidewalk. Our conversation is sparse but it is filled with optimism. No one wants to mention the need for a plan in case something goes wrong and Ana is not released to our custody. We can only focus on the intensity of our faith and the multiple arrangements we must make so we can get her on an airplane bound for Colorado.

We are just a few blocks from the only television station in Bucharest. For years the television broadcast was only allowed to beam images of Nicolae Ceausescu and his destructive propaganda. A few blocks behind us is the city square where gunfire marked the beginning of the end for Ceausescu and his followers.

Ceausescu was out of the country when the December 1989 revolution really began. He made a quick trip back to Bucharest with a grandiose plan to quell the anger that was growing in the streets. His helicopter landed on the rooftop of his primary office building in Bucharest so he could address the crowds. He stepped to his usual ceremonial position on the balcony with the intent of simply reassuring the people and inviting them to all go home. He soon discovered that his words were meaningless. He sensed the overwhelming anger of the crowd. It was a fury that had been building for years. He knew that his only chance was to retreat back to his helicopter for a quick escape. Within moments he was gone but within days he and his wife would be captured, placed on trial and killed by an impromptu firing squad. The world changed for the Romanian people in December 1989. Within just a few hours Ceausescu was dead but the country was in shambles.

Almost five years later the emotional, economic, and spiritual devastation can be seen on each face we pass. No one makes eye contact. Most people carry a bag with the elements of tonight's evening meal. No one smiles, laughs, or even breaks stride as they make their way to their next destination.

In contrast, we can't help but feel hopeful. We are focused on bringing release to Ana but we know there is a chance we will not be successful. If we fail, we will try again. Our eyes will not be diverted from the task. Even the knowledge of our determination is a sign of the internal freedom we feel. No one around us feels the same spirit of internal liberation. Freedom is not something that can be created with documents. It can only be real when it is written on the hearts of the people.

Ana has never known that freedom. She has no internal image of liberation. Her heart has only known the agony of abandonment, the anguish of being alone, and the captivity of the tiny white iron bars that form the boundaries of her world. Her pain motivates us to walk faster.

We have some difficult decisions to make about our travel arrangement. We need to extend our return reservations for a few days but the airline will not guarantee that they will be able to rebook our flights. We have no choice. We must be in Arad for the hearing on Wednesday with the Child Protection Commission. We have Ana's American visa, her Romanian passport, her formal birth certificate, and now we have a German transit visa. We have all the documents, but the only way to actually get her is to let our reservations expire and pray that we can get them renewed when the time comes.

Cheri and I say good-bye to Steve Beard as we hurry out the door on Saturday morning. We had a great time together but he will remain in Bucharest to catch his flight back to the states. Cheri and I will fly back to Arad so we can attend church on Sunday and be ready to work our way through the complexities of the legal system on Monday morning.

I am amazed at the amount of noise one little airplane can make. Twelve people were stuffed into a little propeller driven airplane for a trip to Timisoara. My window seat gives me a chance to notice the black lettering on the leading edge of the wing. It is written in Russian and it has been there for a long time. I hope this plane wasn't a part of the Russian fleet during Stalin's day. It certainly wouldn't qualify as a modern aircraft! The seats are tattered, there is no seat for the single flight attendant, and the pilot looks to be about thirteen years old. Even the atheist former communists on board are praying.

Our prayers had been exhausted and our eardrums were still rumbling when we landed in Timisoara. We taxied to the outer edge of the tarmac before the crew felt comfortable unloading their passengers. Apparently they operate under the principle that walking a great distance while carrying a heavy bag is good for your health. By the time everyone finally gets into the terminal, they have forgotten how bad the flight was. The walk was worse!

About half way to the terminal my ears are assaulted with a loud whistling sound. Three Russian MIG fighters are practicing touch and go landings on the only runway at the airport. Until this moment I had only seen photos of MIG fighters. They look fierce, small, and outdated when compared to contemporary aircraft. It is a good reminder of the heavy military presence.

A bus ride from the terminal to downtown Timisoara lands us at the local taxi station. We inquire with a driver and cut a deal for the thirty-minute drive to Arad. The driver speaks Bulgarian, Romanian, French, and a little German. English is not a language option. It ultimately works out well. We had the opportunity to learn the universal sign language for "I am sorry for the delay, my car is over heating" and "it is a good car but it is very old." The thirty-minute drive took an hour and a half but had a nice conversation on the side of the road in 90-degree weather.

Nelu Redis called Saturday evening to make sure I would be at the church and ready to preach. The service begins at 9:00 AM but we are not expected to arrive until shortly before 10:00. I am to meet Nelu at the back door. He will introduce me to the pastor before the service begins. I still can't believe I am going to speak at the church without having any prior contact with the pastor.

The Saturday night humidity soon became a September Sunday morning swelter. I get dressed, make a few notes about the Lord's Supper, develop a sermon plan, and head down the street to touch base with Nelu and meet the pastor of the Speranta Baptist Church.

Nelu is prompt. He greets me with a cheek-to-cheek kiss and introduces me to several men who are talking at the door. He motions for me to follow him up the stairs, through the door and down the hall. A door with a lace curtain covering the window swings open and I find myself standing with about ten Romanian men all dressed in suits.

They are standing in a circle holding their Bibles in their hands. Their heads are bowed in prayer as first one then another shares a quickly spoken but heart felt prayer. I stand still and listen to the beauty and rhythm of their prayers. The tone and intensity of their mingled prayer rises and falls like the energy of the tide reaching the beachhead after a long absence. It is a potent spiritual moment even though I don't understand a word.

The men lifted their heads as the prayer time came to an end and every eye was fixed on me. I saw warm smiles on every face as one by one they reached across the room to offer the hand of peace. Finally, Nelu stepped forward with a tall young man. He introduced himself in English as Romica. He gave me a nod and a smile as he said, "Some people call me Romi, but the girls call me Romeo." No one needed to translate the joke. All the men laughed as "Romeo" made another joke in Romanian.

This tall handsome young man was the co-pastor of the church. Pastor Romica Iuga and his older brother Viorel serve the Speranta

Church. It is the second largest Protestant church in all of Europe. The wall of the office has a photo display of Billy Graham preaching from their pulpit. One photo shows Dr. Graham walking down the center isle on his way to the pulpit. His stop here in Arad was a part of his 1985 Eastern Europe crusade when he preached the Gospel of Jesus in seven Romanian cities.

In a few moments I will find myself standing behind the same pulpit that Billy Graham used to proclaim Christ. Romi Iuga will stand beside me to translate my halting thoughts as I try my best to say something that will mean something to the people of this magnificent church.

Romi gave me a summary of the service before we entered the sanctuary. There is no bulletin to guide the way. He instructed me to stand when he stands and sit when he sits. He will let me know when it is time to preach. First he will introduce me early in the service so I can bring greetings to the people from my church in Colorado. Then I will have "a few minutes to share something about communion." When the time is right he will let me know that it is time to stand to preach. No problem.

"Don't worry about the sermon time," he whispers as we walk through the sanctuary door, "You will have plenty of time, about forty-five minutes or more if you like."

Forty-five minutes! A mild panic washed across my soul as I looked up to see the 1000 seat sanctuary filled to overflowing. My few notes jotted on a page of my journal might stretch to a full twenty minutes. But I am expected to preach for up to (and even beyond) forty-five minutes. Dr. Lewis, my preaching professor at Asbury Seminary always laughed when he warned us to always be ready to "preach, pray, or die." I didn't really want to die at that moment, so I prayed and then I preached.

I will never forget the experience of sharing my heart with those people. I stood at the massive wooden pulpit and looked out over the congregation. All the women were seated to my left with men filling the section to my right. Most of the women were wearing wool scarves tied around their heads. The children were seated close to the front near the organ. Everyone looked like they were dressed in far too many layers of clothing for the heat of the day. They were attentive and alert in spite of my poorly organized thoughts and the impediment of translation.

Only the choir was mixed in their seating pattern. And what a choir! They sang six different anthems. Nelu directed the choir and selected the various anthems on the spot in response to the message and the focus of the prayer time. He distributed the loose-leaf music sheets and the choir members passed them back as the accompaniment music was beginning to play. The men in the bass section received their music sheets last. They were trying to guess which side of the page they were going to sing as the rest of the choir began to sing. Every piece was a gem. It was a wonderful time of worship.

The Lord used my two favorite passages of Scripture that day. Romans 10:9-13 and Ephesians 4:11-13 provided the basis to remind myself that every Christian is uniquely gifted. Every believer stands under the banner of one God, one goal, one hope, and one Lord. Nations, borders, and political plans cannot separate the heart of God from the people of God.

I was speaking to the people gathered in the sanctuary but I was thinking about Ana. At that moment her crib was less than 500 yards down the street, around a couple corners, and behind an ancient green door. Did she know we were back in town? Did she know that she was a unique gift from God? Did she know we were on the way to change her life?

The worship service went for almost three hours. Time was not an issue. It was a wonderful time of worship. Those moments together launched a long term relationship between my brother Romi, the people of the Speranta Baptist Church, and my personal journey of faith. What a wonderful congregation.

Romi came to our apartment that Sunday afternoon. We talked about the church he served, my church back home, and we talked about Ana. Romi knew parts of our story from previous conversations with Nelu Redis. He was interested in our family situation and listened intently as we told him about Ana. I could see the compassion welling up in his eyes as I told him our story but I could also see a glimpse of confusion. Why work so hard, deplete so much time, and spend so much money on an abandoned HIV+ gypsy girl? It just did not make sense in this culture of struggle and poverty.

Four days later Romi would return to the same living room in the same apartment to meet Ana face to face. He would see her heart and recognize her need. Some of his questions were answered in that moment of personal contact. There are so many things about life that we don't fully understand until personal experience opens our eyes.

Romi wrote his name and address in my journal before he left. I didn't know the impact that first meeting would have on my life. Romi would become one of my best friends and our key contact for continued ministry in Romania. He would help open the doors of opportunity so our combined vision of hope for the children could be expressed in Arad, Romania. In the next few years, Romi would become my teacher, translator, friend, and co-worker for the Lord. We would laugh together, dream of new possibilities for service, deal with the never ending bureaucracy, share our faith, and build a lasting ministry. I will be eternally thankful for the friendship and vision of my brother Romica Iuga.

Ana is waiting

It is Monday morning and we are back on task. We make a trip to the city of Curtici to visit with Ana's birth parents. We are fortunate to find Mr. and Mrs. Cionca in their home. The neighborhood is flooded with barefoot children as soon as our car pulls on to the dirt road that is home to the Cionca family. Everyone in the neighborhood comes running to see what is happening. There are kids everywhere. Most of them look to be under ten years old. They are ragged and dirty with dark matted hair and callused feet. It is my first meeting with the Cionca family and the extended neighborhood clan.

Mr. Cionca is tall and almost distinguished looking. He is obviously poor but he stands proud. He is wearing a tweed sport coat that hasn't seen a cleaning for years. He greets us warmly and points us to Mrs. Cionca who is wiping her hands on her apron. She is short, dark, and has a questioning look on her face.

They recognize Cheri and greet her by kissing her hand and asking about Loredana. They seem to be genuine in their concern for their 10th born child who now lives in America. They respond warmly when Cheri gives them photos of Loredana. Mr. Cionca has a tear in his eye as he shows the photo to the gathered tribe and tells the story of his daughter.

Mr. Black (our translator) begins to speak to momma and papa Cionca in Romanian. He wants to get on with the business at hand so he can get out of this uncomfortable gypsy situation as soon as possible. The Cionca's also know that we are there to talk about Ana and the details of her release. They have already signed waivers and appeared in court to testify that they are willingly relinquishing Ana to our care.

The task today is to invite them to come to Arad on Wednesday to appear before the Child Protection Commission. They will petition the Commission to release Ana to them and then they will release her to our custody. It is the only way Ana can be removed from the orphanage and ultimately leave the country. They agree that it is in her best interest so she can receive medical treatment for the HIV. We make arrangements to have a driver bring them into Arad so we can meet them at the city hall for the hearing.

I had almost forgotten the swarm of children while we were focused on our conversation with Mr. and Mrs. Cionca. Suddenly it is time to go and I realize that a little gypsy boy has his hand in my pocket. No telling what he found. I brush him aside and discover about twenty more children. They are filthy but beautiful. There is no way of knowing what their life is really like. The family has no visible means of support. No one seems to have a job and no one grows a crop. They just live off the dirt.

The kids are crowding in as Mr. Black tries to rescue me by chasing the children away. He tells me to watch my money as he continues to shoo the children with a wave of his hand and the anger of his voice. I remembered that I had a large pack of double mint chewing gum in my backpack and I decided that the kids might enjoy the gift. I sat down in the back seat of the car with the door still open and gave the gum to one of the boys.

I intended for him to share the gum with his friends but he was gone like a streak of lightning and every kid in the neighborhood was hot on his tail. He was running for all he was worth. All the time I could see him taking stick after stick of gum out of the pack and stuffing it into his mouth. If the rest of the pack ever caught up with him at least he would get his fair share before they got the rest. They were still running as our car pulled on to the main road. A moment later, they were out of sight. So much for sharing.

Chapter Nine

New Life from the Breath of God

"Prophesy to these bones and say to them,
"Dry bones, hear the word of the Lord!
This is what the Sovereign Lord says to these bones:
I will make breath enter you, and you will come to life.
I will attach tendons to you and make flesh come upon you and cover
you with skin; I will put breath in you,
and you will come to life.
Then you will know that I am the Lord."
Ezekiel 37:4-6

Wednesday, September 7, 1994

The process for obtaining release for Ana was coming to a close. We still had two major challenges yet to experience. First, we needed to get Ana's birth parents into the county court house to appear before the Commission for Children. One of them must testify that they are giving their permission for Ana to be released from the custody of the orphanage system and that they are in favor of her leaving the country to obtain medical treatment in the United States.

The Commission has responsibility for all of the children who have been abandoned or committed to the child protection system. No one knows for sure, or at least no one will say, how many children are impacted around the country. In fact, no one will say how many children are housed in the system within Arad County. Some believe the number might be as high as 300,000. It is hard to tell. Records are incomplete, orphanages are scattered, and no one wants to loose face by admitting to reality. Either way, the Commission members hold significant influence over the lives of thousands of children.

Mr. Cionca was eager to appear and to share his testimony. He arrived in Arad with his wife and their oldest son, Florin. Florin served a tour of duty in the Romanian military and has enough reading skill to help his parents decipher the necessary documents. He also speaks a little English.

We waited two hours for the secretary of the Commission to call Mr. Cionca for his testimony.We entered the room and faced a long semicircle desk where ten or twelve Commission members were seated. Cheri and I stood beside Mr. Cionca while he prepared for the ordeal. He took off his hat, stood at mock attention, and answered in the affirmative as the Chair of the Commission asked him if he understood the plan to take his daughter to the United States. His response was recorded and he was asked to place his thumbprint on the formal document to make it all legal.

Then the Commission shifted their attention to the two bewildered Americans standing before them. They had three questions. Are we aware that the child Ana Cionca is HIV+? Are we really prepared to take her to the United States for treatment? Will we return Ana to Romania and the custody of the Commission when treatment is complete? Yes, we are aware of her health situation. Yes, we will take her to the United States for treatment. And, yes, we understand that she continues to be connected to the Commission. Everyone in the room knew that Ana would never return to Arad.

The whole process took less than five minutes from start to finish. The Secretary of the Commission smiled, shook our hands, and wished us safe travel as we were escorted out the door. It was over.

One week earlier, the testimony before the Commission looked like a huge roadblock. We had to extend our travel arrangements for a five-minute encounter with the officials. Now it was a thing of the past and we had the paperwork to prove it.

Tomorrow we will bring Mr. and Mrs. Cionca to town again so we can pick up the final documents and obtain Ana's release. By tomorrow afternoon it will all be over. Or will it all be starting?

Thursday, September 8, 1994

Ana's birthday is the 8th of March. It somehow seemed appropriate that the first day of her new life would also begin on the 8th. The day was about to unfold with blurring velocity. The Spirit of God was breathing new life into our lives.

By 9:00 AM we were at the orphanage to visit Ana. The attendants were waiting our arrival with a spirit of anticipation. They were still confused about our motives. Why would people from the United States fall in love with a gypsy girl who is destined to die? Regardless of their questions, they stood ready to help. We would not be allowed to take her in official state owned clothing (top quality that it was) so we brought a full change of clothes for the great escape. Everything fit except her shoes. We had to go back to the market to get sandals that were two sizes smaller. They were still too big!

After a short visit, we met Mr. Black at the City Hall so we could pick up Ana's documents and her official medical history. By 12:30 we had the documents, connected with the Cionca family, and were on our way back to the orphanage.

Mr. Cionca presented the documents to the orphanage director, picked up Ana in his arms, and together we walked out the back door by the kitchen. It was the first time he had even seen Ana since he brought her to the orphanage in 1989. It was the first time he had seen or held his daughter in over five years. The significance of the moment brought tears to his eyes as we made our way toward the green door that led to the street and our waiting car.

Just as we were approaching the green door, my heart skipped when I heard a loud voice shouting from behind. I turned in fear to find the orphanage cook running toward us with her hands wrapped in her white apron. Her short heavy legs were moving as fast as possible as she hurried in our direction. She was crying and talking loudly all at the same time. I didn't understand the words but she was obviously in grief about Ana's departure. She talked, cried and made gestures in Ana's direction. Mr. Black finally eliminated our fears. He translated her words and gave voice to her heart.

She was apologizing. She was responsible for feeding the children but she knew that Ana did not get appropriate food. Her heart was broken by the fact that she often prepared Ana tea water with a crumbled biscuit as her main diet. She knew it was wrong but the food supply was very limited and the other children had a better chance of living. Her tears, her passion, and her broken heart will forever be a part of my life. We said good-bye and assured her that Ana would be fine.

The heavy ancient green door opened with a typical creak to reveal our driver waiting at the street. Mr. Cionca gave Ana a kiss and then handed her to me. Cheri opened the door of the car, got momma and papa Cionca in the back seat and sent the driver on his way to their home. From there we carried Ana the block and a half to our apartment. It was a walk of joy and freedom. Our only fear was that we might fall because Cheri and I were both in tears. It had finally happened. Ana was in our care. She would never ever sleep in a white iron crib again.

Getting out of Town

Our flight to Bucharest was just a few hours away. We were pressed for time but Ana needed a good bath. We drew some warm water, undressed her and began to cry. It was the first time we saw her without several layers of clothing. She weighed just over twenty-four pounds, had no language, she didn't make a sound, she had never eaten solid food, and she was covered with bruises from sitting for hours on the potty board in the orphanage. We washed her with our tears and got her dressed for her journey home. Her tiny fingers were fluttering with the anxiety of a confused child.

Within a few minutes, we had a series of guests. Nelu Redis came with a gift. He brought Ana her very first doll. He was moved deeply by this frail little gypsy girl. Then Pastor Romi Iuga came to say good-bye. He wrote his name and address in my journal so we could stay in touch. He was only with us for a few minutes but he could not take his eyes off Ana. Everyone in the room could see his heart as he began to understand Ana's pain. In that moment be also understood the anguish of children throughout his country. His grief was undeniable.

Sever and Eva Jurca also came. They had been a significant part of this journey. We would stay in touch with them for years. Romi and Nelu would later come to stay with us in our home in Colorado. They will always be good friends.

And then there was Mr. Black. He was our faithful translator. He came to see us off and to wish us well. He only had two requests. First, he wanted us to pray for the health of his wife. Second, if we ever come back, perhaps we could bring him a mystery novel in English. Through the years, we continue to pray for his wife and we continue to supply him with the small treasure of paperback mysteries.

Finally it was time; we had to be at the airport in Timisoara for a flight to Bucharest. By this time, our driver had returned. But this time he had a different car. He was driving a new Ford four door sedan that he borrowed from a friend just for our trip to Timisoara. He said it was his pleasure because he knew this was an important day so he wanted to have a nice car for the trip.

This may have been Ana's first car ride. She was bewildered and obviously wondering what was going on. She was calm and quiet with a quizzical look on her face. She sat directly on my lap facing me as we made our way out of town. Our seating arrangement turned out to be a really bad idea when she got sick and let it all fly while we were sitting face to face in the back seat. That Ford upholstery was never the same.

There we were, just a mile from the airport, stripping down and changing clothes by the side of the road. But no one even turned a head. Before long we were checking in at the airport with a fresh but unique cologne.

September 11, 1994

After two nights in Bucharest, one night in Frankfurt, we were finally on a United Airlines flight to the United States. Ana did great. She became quite a seasoned traveler in a very short time. She loved the window seat so she could check the view.

Our United States customs process and passport connections took place at Dulles Airport in Washington DC. We were hurrying from one station to the next when we were suddenly captured by a wonderful surprise. John, Cheri's husband, had flown from Denver to join us for the last leg. He was all smiles and we were blessed to have him join us for the journey. His original passion for the children of Romania began when he was caught in the destruction of a tornado in eastern Colorado. Now his faith was finding fulfillment through the life of a little girl and the commotion of the airport crowds. Isn't it true, the winds of life blow strongest and best when they are connected to the love of God. John was the face of God in the middle of the turmoil!

We were exhausted but smiling as we boarded our last flight of the day. It was a time of wonderful celebration. We were in American airspace and we were just a few hours from home. Even the airline attendants celebrated with us as we made the last leg into Denver. My heart was racing as we touched down. I knew Sue and our family would be waiting in the terminal.

Colorful balloons and a mob of people were awaiting our arrival at the airport. We were all smiles as we walked off the plane. I was carrying Ana like a proud father walking out of the delivery room. It is fun to give birth to a five year old!

Ana was the center of attention. She was amazed by the whole event. She didn't have any idea that this homecoming celebration was all about her. She just relaxed as everyone came to meet the newest part of our family tree. Aunts, Uncles, cousins, friends, siblings, a full sister, and her adopted mother and grandmother were all there to welcome her home.

Her first day in Colorado called for a continuation of the celebration but it also challenged us to face certain realities. She had no clothes that fit her tiny body so it was time for her first shopping trip. There she was, barefoot in a baby stroller, cruising the aisles looking for an entire wardrobe. The best part of the day came when we found a pair of pink high-topped canvas tennis shoes that would fit those tiny slender feet. She wouldn't let go of those shoes. From that day forward, pink tennis shoes became her trademark.

She was small, thin, and frail. Her health and her ability to survive were in question from the start. She had no language and made no noise. She didn't laugh and she didn't cry. Her first spontaneous giggle was an absolute gift from God.

Her ability to walk was limited by the lack of strength in her legs and a sequence of joints that did not function properly. Years of sitting in a crib, with her legs extended, left her with little flexibility in her ankles, knees, and hips. We carried her from place to place while she learned to maneuver her pink shoes and gained strength in her legs.

We also learned how to do physical therapy as a standard play activity. Every evening her warm bath was another opportunity to massage her ankles and to slowly stretch her Achilles tendons so she could place her foot flat on the ground. Each day Sue went through range of motion exercises with her knees and her hips as they played together on the floor. It was not long before she learned to scoot down stairs on her bottom and eventually gained enough leg strength to walk up a single stair tread.

Ana had never been fed solid food while in the orphanage system. She missed that developmental stage when kids learn to hold solid food in their mouth, maneuver it with the tongue, and chew with their teeth. Swallowing is a whole new world if you have never experienced solid food.

Sue rose to the occasion and spontaneously created a liquid nutritional formula that became Ana's primary diet. But it wasn't that easy. Ana could not (or would not) drink from a cup. So Sue developed a plan to feed her with a 20-millimeter syringe. Just like a little bird, Ana would stretch out in Sue's lap while drop-by-drop she received enough nourishment to survive. There were many days when the process of feeding Ana would take up to eight hours of Sue's day.

Entire days were filled with diaper changes, food preparation, HIV medication, homespun physical therapy, and lots of baby talk to teach a six year old how to say mmmm.

A Day of Celebration

Ana left the orphanage system in Romanian on September 8, 1994. She arrived at her new home in Colorado on September 11th. Years later, the date of September 11th became known for destruction of the New York Trade Center tragedy. For our family, no terrorist act can rob us of the celebration. In fact, Ana not only arrived in the United States on September 11th, but on September 11, 1998, Ana Cionca appeared in Court in Douglas County, Colorado for an adoption proceeding. On that day, Ana was formally adopted into our family. On that day she became Ana Sue Jessen in a Colorado court of law.

This silent child was about to speak volumes to people around the world. Her influence has changed our lives, created ministry opportunities for other children, and reminded thousands of folks that God is still in the miracle business.

After Thoughts

Ana's story is a gift of hope and a testimony to the power of God. She is a daily reminder that faith, hope, and love never fail. As an infant, her health was significantly and tragically compromised by the impact of contaminated blood, archaic medical practices, and the isolation of her childhood.

She is a classic example of what happens when a child is isolated and receives no nurture from those around her. Human contact and interaction is a powerful gift that we can give or withhold. It literally makes all the difference.

Her developmental delays and limited cognitive skills will impact the rest of her life. Nonetheless, her life is a beacon of hope to all who struggle to make sense out of the ordeals of every day life. She continues to break all the rules as she smiles here way through each day. Her wonderful sense of humor forces her to concentrate on ways to tease her mom and dad. It is a great sport!

Ana's HIV status was the most threatening part of her health history. Upon arrival in the United States, medical professionals thought she might live a very short time. Perhaps she would live only six months. Our whole family was ready for that outcome. We were prepared from the beginning to provide her a place in our family where she could know that she was loved to the end. That is still our desire and our commitment.

Thanks to faithful friends, dedicated prayer, and the medical professionals at The Children's Hospital in Denver, Colorado, Ana is strong, healthy, and growing every day. Her HIV is controlled by medication that is constantly monitored and lovingly administered by her mother.

Ana's mom, Sue, not only is the primary care giver, she is teacher, companion, and number one prayer warrior. When Ana was not able to eat, Sue spent as much as eight hours a day holding Ana on her lap so she could feed her a bit at a time from a medicine syringe. For the first several months of her new life, Ana never left Sue's side. Ana woke up with her mom. They spent the day together at Sue's work and it was Sue who tucked her in at night.

It was also Sue who taught Ana to laugh. She became the model of life, the touch of human love, and the voice that brought the gift of words to a silent child.

Today, Ana is our designated family prayer person. At every meal she insists on offering the prayer. It's always the same beautiful prayer. "Dear Lord, thank you for food, Jesus loves me. Amen." Her bedtime prayer still brings tears to my eyes. "Thank you Jesus loves me. Amen."

I thank God for Sue. She is the saint who made it all possible. She literally gave life to a child through an act of love. She helped her eat, taught her to move her knees, and created a home that was filled with laughter and the joy of constant encouragement. She gave Ana the human gift of a loving mother, and even life itself. Some children are simply born in your heart.

We often joke that Ana's health is better than ours. As I write, Ana is about to celebrate her seventeenth birthday and is attending an incredible special education program at Air Academy High School in Colorado Springs. She loves school and has discovered the joy of prom and homecoming. If you ask her what she likes best about school, she is likely to smile broadly and say "boys." As parents, we chuckle that cognitive skills may be delayed but everything else is on track!

The multidisciplinary CHIP Team at Children's Hospital has provided loving care and support that goes above and beyond the challenge of the medical profession.

Our eternal thanks will always go out to the current and former members of the Children's Hospital Infectious Disease Program (CHIP) team. Dr. Myron Levin, Dr. Betsy McFarland, Dr. Adriana

Weinberg, Carol Salbenblatt, Deb Crennen, Janie Kappius, Emily Barr, Robin McEvoy, and Charlotte Stall are just a handful of the team members who gave life to Ana through outstanding long-term care.

When folks thought Ana would never learn to talk, the team at the University of Colorado Speech Language and Communication Center were determined to prove everyone wrong. They started from scratch, researched the possibilities, combined modalities, and lovingly reached for the stars. Today, Ana has an extensive vocabulary and a wonderful sense of humor. The students, instructors, and therapists connected with the CU program loved Ana all the way to effective speech. Sheila Goetz and Cathy Fletcher provided leadership, love and enthusiasm that fueled the learning process.

Our family experience with Ana opened the doors to relationships and ministry that continues to grow in Romania. In 1995 Sue and I took our first mission team to Arad in an effort to share a first hand experience in the orphanage system. That team (and hundreds of team members who followed) gave birth to a ministry that continues to serve abandoned children in Romania.

The ministry of Global Hope was born when people of faith began to believe that they could make a difference. Today, Global Hope owns and operates three homes for abandoned children in Romania. Children come out of the orphanage system and are placed in a Christian group home where they will have a family for a lifetime. The ministry of Global Hope is now beginning to move beyond Romania to meet the needs of abandoned children wherever they are found.

The Global Hope Board of Directors honored Ana by naming one of their group homes after her. Ana's House provides a Christian home for a group of beautiful kids. Log on to www.globalhope.org to see their photos and to gather information about how you can make a difference in the life of a child.

In addition, Ana's ability to overcome the devastating impact of HIV has challenged us to assist other children in Romania who have also been infected with this deadly disease. Romania today has more diagnosed cases of pediatric HIV than any country in Europe. Many more children are already infected but go undiagnosed and thus untreated. Most of those children go untreated because they lack resources or do not have a connection to receive consistent medication. Cases are growing rapidly enough to create fear that Romania will become the epicenter for an extensive pandemic that will devastate large sections of the broader population. Education and quality training for the public and medical professionals is essential to this effort.

StarStone International (the *Star* of Bethlehem and the *Stone* of the empty tomb – everything in between is Jesus) is a non-profit group that has been created to impact the pediatric HIV situation in Romania. StarStone is opening a pediatric diagnostic and treatment clinic in Bucharest to assist in meeting this dramatic need. We dream of the day when StarStone clinics and treatment centers will be scattered throughout Romania (and perhaps the world) in an effort to provide quality care, effective diagnosis, and successful HIV education.

There is one last dream that is coming true for our family. Since Loredana's adoption in 1991 and Ana's arrival in 1994 our daughter and son in law have adopted two other Romanian children. Miamina and Sam are both full siblings of Ana and Loredana.They are beautiful American kids growing up with California sunshine.

Thank God for dreams! Thank God for the Lord of Creation who loves to make dreams come true. Will you join the dream and be a part of the miracle? Today is a great day to make a miracle happen!

Appendix

Our daughter, Ana was born on March 8, 1989 in Curtici, Romania. Shortly after her birth, she was abandoned by her birth parents. She was left in the care of the Romanian State orphanage system in the city of Arad. During her first few days of her life, Ana received a small direct injection of blood. They thought it would give her a boost so she would be healthy. The injection contained blood that was contaminated with the virus that causes AIDS.

Our family spent almost two years trying to create a pathway to bring Ana to the United States. We had one goal, to let her know that she was loved. In September of 1994 Ana entered our family in the United States. The adventure of a lifetime was beginning and we were about to see the church and our society from a whole new point of view.

Along the way we met some very interesting people. We also heard the stories of other people who were raising HIV+ children.

I spoke on the phone with a woman who was very interested in the fact that I serve as a United Methodist pastor. She told me about her experience with a church in our denomination. She wanted the pastor of the church to know about her ten-year-old son. She met privately with the pastor and disclosed to the pastor the fact that her son was HIV+. She wanted to enroll her son in the Sunday School program.

The pastor listened patiently and then outlined a plan that would allow her son to attend the 4th and 5th grade class on Sunday mornings. First, he would inform the classroom teacher that this young man was entering her classroom. He would give her the option of withdrawing as the class teacher. Next, he would inform all the other parents about the situation so they could make a decision about withdrawing their children from Sunday School. Finally, he would inform the Administrative Board so they could be informed of the church's liability.

The mother tried to explain that she did not want her son's health status to be a conflict for the church. She just wanted the pastor to know so he could understand their family situation and provide pastoral care. She asked the pastor to modify his plan but he insisted that it was the only fair option.

The mother told me how she left the church in tears that day. She has not returned to that church or to any other church since that day. Her son has since died. He didn't go to Sunday School and the family didn't receive support from the church. She asked me to explain my colleague's behavior.

At a time when they needed and were seeking the support of the church, they were effectively turned away. Not because the church is bad, but because the pastor was operating out of a decision making process that seemed logical but was actually very painful to that particular family.

Another family made the choice to not tell anyone in the church that their daughter had been diagnosed HIV+. They lived in fear wondering when someone would find out. They feared rejection from their friends and from their pastor. They wanted desperately to tell someone because the load was more than they could handle by themselves. They needed the support of other Christians but it was not available because they simply could not take the risk of disclosing.

Are those unique situations that we can just write off? Perhaps they are experiences from the past before we were enlightened. After all, in this day and age, people are well informed and compassionate. Surely the church would be open to children and families who need the ministry of a caring Christian community. We would pray that it would be so, but the reality is often disturbing.

From our own personal experience, we have discovered that the church is not always prepared for HIV+ children in the Sunday

School, the church nursery, or on the playground. Many people in the church are kind, loving, and compassionate. They don't want to alienate families. But it happens. It happens because pastors, church leaders, and Sunday School teachers have not thought through the implications and options in advance.

We have also faced difficult situations in the church when people were afraid that our daughter would infect their child by being in the same classroom together. We understand the fear and the depth of concern. We trust that solid educational training will solve the problem. Education does not always create a solution. Sometimes fear reigns. Grace must abound.

The goal of this appendix is to provide key information that will help you think honestly about HIV/AIDS. The challenge is to make your church, your home, your heart a safe, welcome, and healthy setting for people with HIV/AIDS. After all, it really isn't "them." It is only "us" or it is no one.

Basic Information

Two sets of letters define our topic, HIV and AIDS. They are familiar to us but we usually don't fully understand the implications until they come to our doorstep. Most people don't recognize how close HIV/AIDS is to your own front door.

HIV (Human Immunodeficiency Virus) was first identified in 1981 when scientists from around the world worked to discover the disease process that was causing the death of young men. Later the term AIDS (Acquired Immune Deficiency Syndrome) became an unwanted part of our human vocabulary. HIV/AIDS is a disease process that destroys the human immune system. When the system has lost its ability to respond effectively, any number of infections and diseases are able to flourish.

Today, there are some wonderful drugs available to people whom are infected with HIV. The newest protease inhibitors have demonstrated a remarkable ability to slow down the duplication of the virus. There is great hope in sight, but the disease process is still fatal. There is no cure.

A variety of medications are available here in the United States. In fact, progress has been productive that we are now seeing a backlash. After a few years of success in keeping the disease under control, the infection rate is now beginning to climb once again. We must double our efforts here in the States while continuing to work faithfully in other parts of the world.

HIV/AIDS is a progressive disease. The virus duplicates and destroys the immune system. The virus actually destroys a cell called the CD4 cell. You may have heard this cell referred to as a "helper cell" or as a "T cell". A blood test can determine the amount of virus that is actually in the system. Another level of blood testing can determine the number of CD4 cells that are remaining in the body. When the level of CD4 cells becomes low, the body cannot fight off infections.

The virus often impacts children dramatically. Depending on the age of the child at the time of infection, their immune system may already be immature. Therefore, they become open to "opportunistic infections" that would typically be dealt with by a healthy immune system.

Testing

There are two basic tests that can determine the HIV status within the blood stream. The ELISA (Enzyme-Linked Immunoabsorbant Assay) blood test can determine the presence of HIV antibodies in the blood. The Western Blot test also identifies the antibodies to the AIDS virus. It is often used to confirm the ELISA. An initial screening can be done with a simple non-invasive mouth swab. All testing is confidential.

Young people, if you have had multiple sex partners over the years and you are planning marriage...please be sure that both you and your fiancé are tested today. Don't wonder about your status. Find out so you can be proactive. You don't want to find out that HIV has been lingering in the background after your child is born and infected.

Transmission

HIV/AIDS is dominantly transmitted through sexual contact or through direct blood contact. Four body fluids contain a large enough concentration of the virus to transmit the disease: blood, semen, breast milk, and vaginal / cervical fluids.

Urine, feces, vomit, and saliva are body fluids that should be treated with care because they may contain traces of blood. Fluids like sweat and tears do not present a hazard.

Today, most children are infected with HIV from their infected mother. Sometimes the transmission of the virus happens during the birth process. Other times the virus is passed along while the child is still in the womb. Many women discover they are infected with HIV after their child develops symptoms. The incubation period for the disease can take several years before symptoms begin to appear in adults. Symptoms often develop more rapidly in children.

An infected mother can be treated with medication during her pregnancy to dramatically reduce the transmission rate. HIV testing is important for every pregnant woman and for every couple thinking about marriage.

Normal daily living experiences and personal contact will not transmit the disease. Hugging, kissing, shaking hands, and general physical contact will not pass the disease from one person to another. Typical activities in any group setting will not transmit the virus.

Sexual contact, shared needles, and prenatal exposure are still the typical means of transmitting the infection from one person to another. The bottom line is this; transmission of HIV is absolutely preventable. It all begins with information that modifies our behavior.

Who is Infected?

Let's face the facts; you have no way of knowing who is infected unless the individual chooses to tell you. Most adults and children will manifest no outward symptoms that would allow you to identify their HIV status.

In fact, many times the infected individuals may not know it themselves! Therefore, we all must assume that all body fluids are infected and treat them accordingly.

Even folks who know they are infected may choose not to tell you. That may seem unfair to you at first. You might even think they owe you the courtesy of informing you. Please remember the power of this disease. Until there is a cure, it will ultimately take the life of the infected person. Remember also, the power of a society that is still poorly informed. That is exactly the reason you must think through the issues now.You cannot wait until someone comes to identify himself or herself as HIV+ and then begin to react. You must pre-think your positive response and prepare your heart for an act of compassion.

Rather than believing that people will gladly will tell you about their HIV status we must approach the issue honestly. We simply must approach every person with respect and compassion. We must treat everyone the same. We must treat all body fluids as being contaminated.

Standard Precautions

The term "standard precautions" (we once used the term "universal precautions) is common language in the medical field today, but is not so well known or understood in the general public. It needs to be a part of our thinking and a part of our language. Every responsible person needs to have a solid understanding of how to use standard precautions.

The use of standard precautions simply means that we treat everyone the same and we approach every body fluid as contaminated. We don't panic when some one bleeds. We pre-think the issues, provide quality training for everyone, and automatically initiate standard precautions.

Do you remember the days when a bloody nose on the playground called for an adult to help hold the child's nose with their hand until the bleeding stopped? Do you remember cleaning up the floor with a rag or a sponge after a bloody nose? Things have changed.

Today, children are taught in school to help care for their own wounds and to get adult help. Adults are trained to place a barrier between themselves and the body fluid. That may mean that the adult takes a moment to slip on a rubber glove (do you have rubber gloves in an accessible location?). Sometimes a rubber glove may not be available. In that case a towel or a newspaper or any other item that creates an effective barrier can be used. Do not use your bare hand to catch blood or to clean up body fluids.

Effective education in the area of using standard precautions will help everyone be safe and it will eliminate the possibility that any one individual is singled out for special precautions. Any time we focus on the "infected" person we will be isolating that individual. We may even become guilty of driving them and their families away from one of their most valuable support networks.

Using standard precautions basically comes down to three items:

1. *Treat everyone the same and all body fluids as infected.*
2. *Create a barrier between you and the body fluid.*
3. *Dispose of contaminated items appropriately. Double bag rubber gloves, towels, or other items that have come in contact with body fluids.*

Teach standard precautions to everyone. It is the right thing to do for your students and for your staff. Be sure every leader is trained and that rubber gloves are easily obtained.

You might also be faced with a situation where a coworker, a teacher at school, a childcare worker, or a church leader is infected with HIV. The same standard precautions that protect you from infection will also provide protection for your children. Remember HIV is not transmitted by being in the same room or by having normal physical contact with the individual. You may actually be more of a threat to their health than they are to you.

This is a wonderful area to call upon the medical personnel. Invite a physician or a nurse to be a part of your neighborhood gathering. Have them describe the value of using standard precautions. Help people understand how easy it is to protect everyone from a variety of unwanted infections.

Sometimes people in the community can hear and understand the medical value of using standard precautions from one of their own members. They have a built-in trust network. Other times the message may be heard with more power if you bring in a special guest as an "expert" to talk about this particular topic.

Legal Issues

Let's start this section with a basic principle. **The no person should ever be forced by law to do what is morally and ethically right**. There are legal issues we all need understand, but the power of our faith should lead us to live ahead of the law.

Some laws regarding HIV/AIDS have been passed to prevent discrimination. We cannot pass judgment on any person because of their health situation. People, especially children and their families, need the support of trusted friends. We simply cannot turn them away even if we are not comfortable with their situation. It is our responsibility to deal with our discomfort.

Issues related to confidentiality sometimes create a moral dilemma. Infected individuals of all ages have the right to disclose or not disclose their HIV status. No one else has the moral or ethical right to disclose another person's HIV status without permission.

Ask yourself some serious questions. Do you have the right or the responsibility to share information about a person who is HIV+? No. Do people have the right to expect confidentiality to be maintained? Yes. Regardless of age or personal situation, we all have a moral responsibility to maintain confidentiality.

Here is a five-step plan for making an impact right where you are.

1. Begin with your own level of information. Read an up to date book about HIV/AIDS, talk with your physician, and check with the county health department. Do whatever you need to do to inform yourself and to be open so you can help others.

2. Schedule a conversation with your family. Be intentional about it. Don't wait until a critical issue arises. Be sure to talk about how HIV is transmitted and how it is not transmitted. If you have teenaged children, start talking about appropriate sexual behavior. Begin to teach the people you love so they can talk about their feelings and be ready for every day life.

3. Move your family conversation from talk to action. Be sure every person knows and practices Standard Precautions. Have a practice session that is followed by conversation and evaluation.

4. Create an information sharing task group in your community, through your service club, or in your place of worship to continue the education process with a broader circle.

5. Create a plan to discover what is really happening on the global scene. Don't depend on the television news or even on your local newspaper. Get a copy of Don Messer's book *Breaking The Conspiracy of Silence*. Read it, take it to heart, and find a way to help others understand the magnitude of the problem. Then find a way to get involved at a hands-on level. There is much that needs to be done. Log on to the Baylor University HIV/AIDS web site www.bayloraids.org for direct information and top quality training materials. Check www.starstone.org for information about how you can help support international HIV/AIDS ministry.

Whatever you do, please don't assume that the people around you know all they need to know about the subject. And above all, don't get caught in the trap of thinking it is someone else's problem. People make a difference. People need you! HIV/AIDS is never about "good" people and "bad " people. It is about people. And people need people, especially when the days are difficult.

Printed in the United States
54120LVS00002B/340-342

9 781420 891690